MATTER, ENERGY, AND RADIATION

MATTER, ENERGY, AND RADIATION

A Syllabus for Science A1, the First Semester of a Two-Year Course in Science in Columbia College

J. R. DUNNING
H. W. FARWELL

NEW YORK: MORNINGSIDE HEIGHTS
COLUMBIA UNIVERSITY PRESS
1935

COPYRIGHT 1935
COLUMBIA UNIVERSITY PRESS

PUBLISHED 1935

LITHOGRAPHED IN THE UNITED STATES OF AMERICA

Preface

This Syllabus has been prepared for students in Science A1, the first semester of a two-year sequence in the sciences in Columbia College. The course Science A-B is designed for those students who do not intend to use the sciences professionally, but who desire a general acquaintance with the chief fields of scientific investigation - a discussion of their dominant problems, concepts, and theories, and an introduction to the techniques of experimental methods. Its aim is to present as systematically as possible those themes of modern science that are of general interest and significance. The opportunity is open for the student to go as far as he wishes. The course is conceived as a unified program of study for students who desire to satisfy the science requirement for the Bachelor of Arts degree. The major topics of the two-year program are shown on the following pages.

This Syllabus for Science A1 is not intended to serve as a text, but rather as a guide to the students' thinking, focusing their attention on the three topics: MATTER, ENERGY, and RADIATION, emphasizing the important sub-topics, and opening up the more vital questions in the various fields, rather than going into the smaller details.

There is no definite textbook for Science A1, although Lemon, "From Galileo to Cosmic Rays," and Loeb and Adams, "Development of Physical Thought" are used as the principal reference texts. In order to take advantage of the opportunities offered by the course, the student is urged to read widely from the list of Essential and Suggested Readings, placed at the end of each week's section of the Syllabus. The special collection of books and periodicals in the Science Reading Room is shown at the end of the Syllabus.

For the laboratory the class is divided into as small sections as possible. In each laboratory period an attempt is made to cover a number of significant experiments, some of which are quantitative and some of a qualitative nature. In some cases the experiments are performed individually, but in many cases they are performed by the group as a whole under the guidance of the instructor. This method makes possible an experience for the student which the conventional individual experiment cannot afford. The laboratory work is supplemented by visits to the Hayden Planetarium, the Rutherfurd Observatory, and the various research laboratories in this and other departments. An opportunity for a certain amount of special individual work will be provided.

The suggestions and assistance of the other members of the committee in charge of the science course, and of the members of our own department, have been very valuable in preparation of this Syllabus.

J. R. Dunning
H. W. Farwell

Pupin Physics Laboratories
Columbia University

August, 1935

Contents

General Outline of Two-Year Science Course. ix

First Week. Section I. <u>GENERAL INTRODUCTION</u> 1

Second Week. Section II. <u>MATTER</u> 5
    Three States of Matter.
    Properties of Matter: Mechanical, electrical, etc.

Third Week. Section II. <u>MATTER</u> (continued). 8
    Physical and Chemical Properties.
    Atomic Theory.

Fourth Week. Section III. <u>ENERGY</u>. 11
    Energy - a Purchasable Commodity.
    Mechanical Energy.

Fifth Week. Section III. <u>ENERGY</u> (continued) 14
    Principle of Conservation of Energy.

Sixth Week. Section III. <u>ENERGY</u> (continued) 18
    Electrical Energy - Production and Conversion.
    Conservation of Energy and Transformations.

Seventh Week. Section III. <u>ENERGY</u> (continued) 22
    Particle Nature of Matter.
    Kinetic Theory of Matter - Gases.

Eighth Week. Section III. <u>ENERGY</u> (continued). 27
    Kinetic Theory - the Gas Laws.
    Gases, Liquids, and Solids.

Ninth Week. Section III. <u>ENERGY</u> (continued) 32
    Kinetic Theory.
    Continuity of State.
    Thermal Cycles. Structure in Matter.

CONTENTS

Tenth Week. Section IV. <u>RADIATION</u> 37
    The Electron - Its Properties.
    Electrical Circuits.
    Conduction in Gases.

Eleventh Week. Section IV. <u>RADIATION</u> (continued). 43
    The Electron. Photoelectric Effect.
    Thermionic Emission.
    Electromagnetic Radiation - Radio.

Twelfth Week. Section IV. <u>RADIATION</u> (continued) 49
    Temperature Radiation - Quantum Theory.
    Radiation from Excited Atoms.
    Spectroscopy.

Thirteenth Week. Section IV. <u>RADIATION</u> (continued). 53
    X-rays - Crystal Analysis.
    Electromagnetic Radiation.

Fourteenth Week. Section IV. <u>RADIATION</u> (continued). 57
    The Nucleus. Radioactivity.

Science Reading Room and Book Collection. 63

General Outline of the Course Science A-B

In order to provide a comprehensive view of the scope of the whole course, these four brief outlines for Science A1, A2, B1, B2 indicate the major topics of the four semesters.

SCIENCE A1

MATTER, ENERGY, AND RADIATION

I. **GENERAL INTRODUCTION**

    Science and civilization.
    What do we mean by science?
    The importance of experiment.

II. **MATTER**

    The three states of matter - solid, liquid, gas.
        State is an accident of temperature and pressure.
    Physical properties.
        Density, mechanical, thermal, optical, electrical, etc.
    Physical and chemical phenomena.
        Elements, compounds, and mixtures.
    The atomic theory.
        General atomic structure, nucleus, external electrons.

III. **ENERGY**

    The concept and measurement of energy. A purchasable commodity.
        Mechanical energy. Force and distance.
        Potential and kinetic energy. Energy conversions.
    Principle of Conservation of Energy.
        Application to all forms of energy - heat, electrical, etc.
    Transformation of energy into various forms.
        The basis of modern civilization.
    The particle nature of matter.
        Atomic structure and the Periodic Table of Elements.
    The kinetic theory of matter. (Gases, liquids, and solids.)
    Structure in matter - crystalline and non-crystalline.
    Statistical nature of kinetic theory and theories in general.

IV. **RADIATION**

    The electron.
        Electrical circuits, discharge in gases, P.E., etc.
        Thermionic emission, electromagnetic radiation.
    Temperature radiation. Planck's quantum hypothesis.
    Spectra.
        Spectral lines and series. The Rutherford-Bohr atom.
        The quantum theory.
    X-rays - crystal analysis - atomic structure of matter.
    Electromagnetic radiations: Radio - cosmic rays.
        The particle-wave controversy.
    The nucleus - a new era in physics.
        Isotopes and radioactivity.
        Nuclear transformations, and nuclear energy.
    The equivalence of matter, energy, and radiation.

GENERAL OUTLINE OF SCIENCE A-B

SCIENCE A2

CHEMICAL CHANGES IN MATTER

I. **ELEMENTS, COMPOUNDS, and MIXTURES**

   The elements and chemical reactions.
   Oxygen, hydrogen, and water.
   Atoms and molecules - reviewed.
   Chemical formulae, equations, valence.
   The atmosphere and its components.
   Use of the Periodic Table of the Elements.

II. **SOLUTIONS**

   Concentration and saturation, pressure and temperature.
   Ionization.
      Acids and bases. Neutralization.
      Electron transfer. Oxidation and reduction.
   Colloidal solutions.

III. **CHEMICAL EQUILIBRIUM**

   Rates of reactions. Reversible reactions.
   Nitrogen fixation.

IV. **SOME GROUPS OF THE ELEMENTS**

   Halogens, sulfur, phosphorus.

V. **THE METALS**

   Alkalies and alkaline earths.
   Aluminum and iron. Metallurgy.
   Electrochemistry.
   Alloys - other metals.

VI. **CARBON COMPOUNDS**

   The carbon atom, structural formulae, isomerism.
   Hydrocarbons. Combustion.
   Carbohydrates and photosynthesis.
   Alcohols, acids, and esters. Fermentation.
   Fats and proteins. Putrefaction.

VII. **FOODS**

   Nutrition. The vitamins.
   Introduction to biological methods in chemistry.

SCIENCE B1

THE EARTH, ITS ORIGIN AND PHYSICAL HISTORY

I. INTRODUCTION

II. ASTRONOMICAL ASPECTS

    The universe. The galaxy. The solar system.

III. THE EARTH AS A PLANET

    General facts.
    Time.
    The moon.
    Map projections.

IV. THE EARTH AS A GEOLOGICAL BODY

    The crust of the earth.
    Minerals.
    Rocks.

V. GEOLOGICAL PROCESSES

    Weathering. Streams. Underground processes.
    Glaciers. Waves. Wind.

VI. GEOLOGICAL STRUCTURES

    Horizontal structures.
    Domed structures.
    Faulted structures.
    Folded structures.
    Complex structures.

VII. REGIONAL GEOLOGY

    The United States.

VIII. HISTORICAL GEOLOGY

    The geological column.
    Evolution.
    Pre-Cambrian geology.
    The Paleozoic.
    The Mesozoic.
    The Cenozoic.
    The evolution of man.

GENERAL OUTLINE OF SCIENCE A-B

SCIENCE B2

TRANSFORMATIONS OF MATTER AND ENERGY IN LIVING ORGANISMS

I. **INTRODUCTION**

    Survey of plants and animals.
    Similarities in structure and function.

II. **LIVING MATTER**

    Chemical composition. Elements - compounds.
    Dynamic constituents - Enzymes.
    Basic processes. Diffusion, osmosis, catalysis.
    Sources of materials in nature.

III. **SOURCES OF ENERGY IN LIVING ORGANISMS**

    The green plant.
        Structure. Photosynthesis - osmosis, enzyme action.
    Growth - reproduction, heredity, and evolution in plants.

IV. **UTILIZATION OF ENERGY IN MAINTENANCE**

    Metabolism.
        Digestion, distribution.
        Respiration - excretion.
        Food and energy cycles.
    Sensory activity.
        Coordination - movement.
        Stimulus and response.
        The adjustory system - nerves.

V. **UTILIZATION OF ENERGY IN ADVANCING PROCESSES**

    Growth.
        Cell division.
        Embryology.
        Physiology of development.
    Reproduction.
        Asexual and vegetative.
        Sexual - chromosomes.
        Reproductive organs.
    Heredity and evolution.
        Mendel's law.
        Heredity and environment.

VI. **INTERDEPENDENCE OF PLANTS AND ANIMALS**

    Food cycles.
    Bacteria.
    Ecology.

VII. **HUMAN BIOLOGY**

First Week

SECTION I - GENERAL INTRODUCTION

A. SCIENCE AND CIVILIZATION

Why is science so important for the educated man of today?
   In what respect is science valuable to men who are not to be professional scientists?
      In subject matter, as such? In method? Or in both?
   No matter what field you enter, why can you be certain that the past and future developments in science will greatly influence your work and thought?

What is the fundamental idea behind this offering of a two-year science program?
   The program for the four semesters, whose major topics are shown on pages ix-xii, offers you an opportunity to become acquainted with the chief fields of scientific investigation, with their dominant problems, concepts, and theories, and with the techniques of their experimental methods.

What are the sciences, and how did the arbitrary distinctions arise?
   Was Galileo a physicist or an astronomer?
   Was Newton a mathematician or a physicist?
   Is the trend toward further specialization and sharply separated fields in science, or is there a trend toward a realization that the same basic principles underlie all the sciences?

If there has been a development of our civilization, wherein has science played a part?
   Is science likely to become a still more important factor in the future, and exert a more profound influence on our social, industrial, and political order, or has science reached the peak of its influence?

Such questions as these can hardly be answered by any one man, but in the consideration of them we can acquire a greater understanding of the world in which we live, the circumstances which have led to our present conditions, of our contemporary problems, and of some of the possibilities which the future offers.

B. WHAT DO WE MEAN BY SCIENCE?

Before we go far into any one field we ought to have some understanding of this question.
   In a sense, science is a product of man's curiosity, of his imagination, and often of his pressing need.
   Man's desire to understand the nature of the world around him, his long search for ideas and for truth have not produced merely a dead set of facts, but a living, fascinating drama which has really only begun.

FIRST WEEK

B. CONTINUED

Arbitrary definitions are rather dangerous, and we may understand what we mean by science better through considering the way in which a science comes into existence and grows. For example:
  How did the earliest machines come about?
  How did geometry develop from the "rope stretchers" in Egypt?
  Better still, consider the gradual development of our ideas of the solar system, beginning with man's first wonderment at day and night, through the development of early and modern astronomy, on up to our current notions about expanding universes.

<u>Rise of astronomical knowledge</u>. An example of the growth of science.
  Early superstitions. What ideas did the cave men have?
  Beginning of observation.
    What part did religion, the early priests and astrologers play in beginning experimental observation?
    How far did the early civilizations, such as the Babylonian, go?
  The Greek School.
    Why has the long line of Greek thinkers from Thales and Pythagoras on down to Aristotle had such an influence?
    What led to the general acceptance of the geocentric theory?
  The interim of seventeen centuries.
    What factors played a part in arresting the growth of science during the seventeen centuries from the Greek School to Copernicus?
    Are such factors still operating in the world today?
  The beginning of modern science.
    Why has the development of the <u>experimental method</u> through the work of Copernicus, Galileo, Tycho Brahe, and others been so important, and had so many repercussions in other fields of human activity?
    What are the basic facts of the solar system?
      How does the work of Tycho Brahe, Kepler, and Newton illustrate the effectiveness of a combination of theory and experiment?
      The planets, Mercury, Venus, Earth, Mars, Jupiter, Saturn, Uranus, Neptune (with their various satellites or moons) traverse elliptical orbits, with the sun at one of the foci, the ellipses being nearly in the same plane. - Kepler.
      The distances range from about 36 million miles for Mercury, to about 2790 million miles for Neptune.
      The Earth is about 93 million miles from the sun, has a period of about 365 1/4 days, and a radius of about 4000 miles, while its satellite, the moon, is about 240,000 miles distant and has a period of about 27.3 days.
      The squares of the times of revolution of the planets are proportional to the cubes of their mean distances from the sun. $T^2 \sim r^3$. - Kepler.
      The straight line joining the sun to a given planet, (radius vector), moves so it sweeps out equal areas in equal periods of time. - Kepler.
      The planets move in equilibrium between the gravitational attraction force $F = kmm'/r^2$ toward the sun and the

FIRST WEEK

B. CONTINUED

centrifugal force outwards, $F = mv^2/r$.
These concepts and developments from them will be considered more fully in Science B1.
In what ways does the work in the fields of physics, chemistry, and mathematics, by Newton, Leibnitz, Boyle, Huyghens, Laplace, and others, in its reaction on astronomy, show the interdependence of all fields of science?

Modern astronomy and astrophysics.
Our growing picture of the universe - the stars, galaxies, nebulae - expanding universes, relativity.
Hale, Hubble, Tolman, Einstein, LeMaitre, and others.
The future - our knowledge of the universe only beginning.
The cooperation of physics, chemistry, geology, mathematics.
Does this development of astronomy consist simply of ever-increasing accumulation of facts?
How does this brief survey of one science help us understand what we mean by science?

C. THE RÔLE OF EXPERIMENT

What is an experiment?
Why does the substitution of controlled quantitative experiments, for wild ideas partly or wholly unrelated to facts and for frequently misleading hit-or-miss happenings, represent such a great step forward?
To what extent have we actually taken advantage of the scientific method in other contemporary human undertakings?
How do theories grow out of the results of experiment? What is the rôle of theory in science?
Why must an accepted theory rest fundamentally on results of experiment?
How do experiment and theory go hand in hand in much of our scientific work?
In what ways do the possibilities of experimentation differ widely in the various sciences?

IN THE LABORATORY

Special Visit to Hayden Planetarium
As an introduction to science, Dr. Clyde Fisher of the American Museum of Natural History has arranged a special program for us at the Hayden Planetarium, 77th Street and Central Park West. The dates for this visit will be announced.
What types of phenomena connected with the solar system and the universe is the Zeiss Planetarium projector capable of showing?

Rutherfurd Observatory.   Regular Laboratory Period
At the first regular laboratory meeting Professor Schilt and the Astronomy Department have provided an opportunity to visit the Rutherfurd Observatory on top of the Pupin Physics Laboratories, in order to study the equipment of the observatory, the large refractor telescopes, with equatorial mountings, the transits, and the various other tools of the astronomer.

FIRST WEEK

On one or more evenings during the week (to be announced) we have been invited, weather permitting, to view some of the stars and planets through the 12-inch refractor.

FOR STUDY AND READING

ESSENTIAL
Loeb and Adams, "Development of Physical Thought," pp. 1-54.
Or Knowlton, "Physics for College Students," pp. 1-11, 65-73.
Or Lemon, "From Galileo to Cosmic Rays," pp. 3-8, 12-31.

SUGGESTED
Fath, "The Elements of Astronomy."
Lenard, "Great Men of Science."
Lodge, "Pioneers of Science," chs. 3, 4, 5, on Galileo.
Mayer, "Seven Seals of Science," p. 211.
Newman (Moulton), "Nature of the World and Man."
Sedgwick-Tyler, "Short History of Science," p. 191.

Second Week

SECTION II - MATTER

A. **INTRODUCTION**

All sciences primarily study matter in its various forms and relationships, and nearly all human activities have definite dependence on matter and its properties. Hence we ought first to find out as much as possible about our working materials.

Matter may exist in <u>three fundamental states</u>, gas, liquid, or solid.
  Can all substances exist in all of the three states?
  How do temperature and pressure determine the state of any given substance?
  With what advantages in this respect have our accidental circumstances provided us? Suppose our average temperature were 30°C lower? 30°C higher?
    How much of a change in average temperature would it take to cause another ice age?
    How much do we know about the behavior of matter at the lowest temperatures obtainable?

B. **PROPERTIES OF MATTER**

Let us consider some of the typical properties which make it possible to distinguish one kind of matter from another.
  <u>Density</u> differences are quite marked.
    Density = Weight per unit volume = W/V.
    How are the basic units of C.G.S. (centimeter-gram-second) system defined and used?
    Why does the scientist need <u>exact</u> definitions?
  Mechanical Properties.
    Elasticity, malleability, ductility, hardness, strength, viscosity, etc. provide a basis for distinction.
      Are all of these properties found in all states of matter?
      Do all substances follow Hooke's law, $F = -kx$?
    How much does industrial progress depend on the mechanical properties of matter, and the development of special materials with special properties?
      What about aluminum, stainless steel, rubber, bakelite?
      What other special materials are needed today?
  Structure in Matter.
    Large differences in structure exist, from the crystalline to the so-called non-crystalline forms, but actually all matter has structure.
    How small can we subdivide a piece of matter and still have it preserve its characteristics?
  Thermal Properties.
    Such thermal properties as conductivity, melting point, boiling point, and expansion with change of temperature are highly important.
      Can we improve on the thermal properties of materials found in nature? Heat insulators?

SECOND WEEK

B. CONTINUED

The boiling points range all the way from -269°C for helium to 5900°C for tungsten.

Optical Properties.

The index of refraction, the transparency and color of materials are involved in fundamental problems.

Are each of these properties independent? For example, does the index of refraction depend on color?

Electrical Properties.

The electrical conductivity, for example, varies from nearly zero for such materials as amber, to nearly infinity for superconducting metals at very low temperatures.

Would you expect electrical and thermal conductivities to be related?

Does the fact that the conductivity of a material is a constant independent of the current flowing (Ohm's law $R = V/I$) seem fortuitous?

This is an incomplete, but typical, list of properties, some general, some specific, which enable us to describe matter.

In order to understand these properties of matter in bulk, we must go deeper, to the particles of which matter is built, the molecules, the atoms themselves, and even the electrons and the other component parts of the atoms.

IN THE LABORATORY

How can we measure some of these typical properties of various materials?

Density.

In order to become familiar with some typical materials in nature, and methods of measurement, make use of simple methods based on correct definitions to determine the density W/V of a wide range of materials:

Gases: CO_2, air (N_2 and O_2) and H or illuminating gas; various liquids and solids.

Can you suggest other methods of measuring density which would be applicable?

Determine as many other properties of the substances available, such as color, hardness, etc. which individual inspection will disclose.

Mechanical Properties.

Test the mechanical properties such as hardness, flexibility, breaking strength, etc. of a number of materials with the apparatus available.

From your measurement of the tensile strength of steel wires compute the breaking strength of one of the 36-inch cables of the George Washington bridge.

Structure.

Examine directly, and under microscopes with various magnifications, a sufficient variety of samples to learn something about crystalline structure.

SECOND WEEK

Do all of the materials examined have structure?
What forms can you definitely identify?

FOR STUDY AND READING

ESSENTIAL
    Lemon, pp. 91-105, 109-115, 177-184, 191-196, 265-268.
    Or Loeb and Adams, pp. 74-90, 198-206, 317-322, 418-421.

SUGGESTED
    Joffe, "Physics of Crystals."
    Knowlton, pp. 131-142.
    Saunders, pp. 305-310.

Third Week

SECTION II - MATTER (Continued)

C. PHYSICAL AND CHEMICAL PROPERTIES

Why do we distinguish between <u>physical</u> and <u>chemical</u> phenomena?
   Is this distinction real, or somewhat arbitrary?
   Should the physicist or the chemist study the conduction of electricity through metals? Through solutions? Through gases?

With what properties of matter is the chemist chiefly concerned?

How do we know whether a given lump of material is a <u>mixture</u>, a <u>compound</u>, or <u>an element</u>?
   Is the definition of an element given by Boyle in the "Sceptical Chymist" satisfactory now?
   Does the concept of atoms seem essential in describing chemical processes involving the changes in composition and energy which occur when elements interact with each other?

D. THE ATOM

Should we call the early suggestions of Democritus and Leucippus, that matter is composed of elementary particles, more than pure speculation?

Why was the general acceptance of the atomic nature of matter delayed until after the discovery of radioactivity, by Becquerel, in 1896, even though the earlier experimental work of Dalton, Gay Lussac, and Avogadro on gases had clearly indicated the existence of atoms?

What do we mean by an <u>atom</u>?
   Is the word etymologically correct now?
   Are single atoms (about 10^{-8} cm in diameter) visible under a microscope? Can we prove by other methods that atoms really exist?
   If it is so small, why should we be able to detect by electrical methods the nucleus of a single helium atom (alpha particle) when it is ejected from radioactive elements?

The 93 major kinds of atoms, or <u>elements</u>.
   Are some still unknown?
   Note the arrangement in the Periodic Table of the elements, of which more later.
   In what respects do these 93 types of atoms differ? For an example, how is an iron atom (Fe) different from an aluminum atom?
   How much do we know about the internal structure of atoms of the various elements?

<u>Simple Concepts of Atomic Structure.</u>
   Roughly analogous to the solar system.
   <u>The simple H (hydrogen) atom</u>, about 10^{-8} cm diameter.
      <u>Central nucleus</u> - proton - mass about 1.007 units.
         1.66×10^{-24} gm, about 10^{-13} cm "radius."
         single + charge.

THIRD WEEK

D. CONTINUED

    External "planetary" electron, mass 1/1840 proton.
      single - charge.
    A *dynamic atom* mostly empty space.
  Atomic number = Number of external electrons = no. protons in nucleus, to be electrically balanced.
    Atomic number of H = 1, of U (uranium) = 92.
  Atomic weight: weight of atom (largely due to nucleus) on arbitrary scale, with oxygen^{16} = 16.0000.
    Atomic weight of H = 1.007, of U = 238.

Why must all the properties of matter, physical, chemical, biological, etc. depend ultimately on the characteristics of these atoms and their constituents?

How do we make chemical combinations of these atoms?
  Do they all form combinations?
  Do atoms differ in the way they form combinations with other kinds of atoms?
  Molecule: A combination of two or more atoms of the same kind such as O_2, N_2, etc., or of atoms of a different kind, such as CO_2, H_2O, etc.
  Are ordinary physical processes violations of the Principle of Conservation of Matter, i.e., that matter can neither be created nor destroyed?
    What about ordinary chemical transformations?

Have we any evidence to justify the saying that all matter is fundamentally electrical in nature?

IN THE LABORATORY

Physical Mixtures.
  What is a *physical mixture*? What methods are available for the separation of such mixtures?
  Study the effect of gravitational force in causing mixtures of fine silts, clays, sands, etc. to settle out of water mixtures.
    Does *Stokes' law*, that the rate of fall of a particle is proportional to the square of the radius of the particle divided by the viscosity of the fluid, seem reasonable?
    Why should the finer particles, as in the case of the gold colloid, remain suspended indefinitely? Is this fact connected with the constant motion of the water molecules?

Centrifuges.
  Observe the action of the high speed centrifuge in greatly increasing the rate of settling of the particles. What sets a limit to this method?
  If the centrifugal force $F = (W/g)(v^2/r)$, how much greater than the gravitational force W is the force F on the particles in the centrifuge? The centrifuge revolves at about 4000 RPM, and r = 20 cm.

See how effective purely physical methods are in separating a physical mixture of various materials, such as sand, lead shot, iron filings, and salt.

THIRD WEEK

Are separations in general completely <u>quantitative</u>? Or are they always <u>fractional</u>?

Could we use chemical methods of separation more effectively?

<u>Detecting Single Atoms</u>.
    After "dark adapting" your eyes, observe the tiny scintillations produced in the zinc sulfide (ZnS) screen of the "spinthariscope," and the various microscopes by the individual alpha particles (nuclei of helium atoms) which are ejected from radium or radium F (polonium).

        The velocity of alpha particles from polonium (radium F) (about 12,000 miles per second) is such that in spite of their tiny weight of about 6.6×10^{-24} gms, each single alpha particle has sufficient energy to disturb a molecule of zinc sulfide to such an extent that it produces a tiny burst of light large enough to affect the human eye.

    Do these scintillations occur regularly, or at random?

    What makes your watch dial glow in the dark? Observe it under the microscope.

FOR STUDY AND READING

ESSENTIAL
    Bazzoni, "Energy and Matter."
    McPherson and Henderson, "General Chemistry," pp. 1-26.
    Or Loeb and Adams, pp. 245-260.
    Or Lemon, pp. 316-325.

SUGGESTED
    Crowther, "Ions, Electrons, and Ionizing Radiations."
    Jauncey, "Modern Physics," pp. 358-363.
    Mayer, "Seven Seals of Science," p. 239.
    Rutherford, Chadwick and Ellis, "Radiations from Radioactive Substances," pp. 54-58, 544.

Fourth Week

SECTION III - ENERGY

A. INTRODUCTION

What is *energy*?
   What primarily do we buy when we purchase food, electricity, coal, gasoline, or gas?
   Is energy the most fundamental basis of value?
       How much of human activities can be put purely on an energy basis?
       What about energy as a medium of exchange?
   Since energy is so important and its utilization the foundation of progress and of life, we ought to have a thorough understanding of its nature.

B. IDEAS ABOUT ENERGY

MECHANICAL ENERGY

Why did the concept of *mechanical energy* develop first?
How do we use the expression:
   Energy = Force x distance (grams x centimeters or lb x ft) to measure energy or work?
Must the force be in the same direction as the motion?
   If a force of 150 lb is required to move a 3200-lb car a distance of 20 ft, how much energy has been utilized?
Are the units *foot-pounds*, or *gram-centimeters* adequate to measure all mechanical energy?
Why is the physicist so interested in "units" or "dimensions"?
What about other types of forces, such as electrical?

POTENTIAL ENERGY

With what examples of *potential energy*, energy stored up, do you come in contact in everyday life?
Is the expression:
   Potential energy = W x h (gm x cm or lb x ft) adequate?
Since gravitational potential energy depends on the height h, it is always "relative." *What about other forms?*
A certain gasoline advertises that a gallon of it will lift a 3000-ton locomotive 15 ft. How much potential energy per gallon does this indicate? In what form is it?

KINETIC ENERGY

How must we use the relation:
   Kinetic energy = $1/2\ W/g\ v^2$ to measure the energy of moving bodies?
Does kinetic energy have the same units, gm cm or ft lb, as potential energy?
   How are the quantities that enter into the expression defined, and what units do they have?
The quantity W/g, i.e., weight/acceleration of gravity, is often called the *Mass*. $g = 980\ cm/sec^2$, or $32\ ft/sec^2$.

FOURTH WEEK

B. CONTINUED

Conversion of Mechanical Energy.

Mechanical to potential energy, or vice versa.

What happens when a book falls from the table?

Are we justified in believing that all of the potential energy of a body relative to the ground is converted into kinetic energy if it falls to the ground?

Cars have occasionally skidded off the 125th Street viaduct in icy or wet weather, and plunged about 75 ft to the street below. How much potential energy did a 4,000-lb car have before the plunge just as it was poised on the edge? How much kinetic energy did it have as it hit the street? What happened to the kinetic energy?

How much of the mechanical work in industry is based on conversion of potential to kinetic energy?

Where does the energy come from originally?

IN THE LABORATORY

How can we measure energy, directly or indirectly, in actual practice?

What methods do we have for measuring _force_?

Does the stretching of a spring obey Hooke's law, $F = -kx$, sufficiently well so that it may be used accurately to measure force?

What about the beam balance, or platform balance?

Why do we use _machines_?

Investigate what happens to the energy in a number of different machines, such as a pulley system used to lift a heavy weight, an inclined plane, a differential pulley, a gear system, or automobile transmission.

Measure the energy output or work done
   i.e., Load x distance load moves (ft.lb or gm.cm)
Measure the energy input
   i.e., Force x distance force moves (ft.lb or gm.cm)
Efficiency = $\frac{\text{Output energy}}{\text{Input energy}}$

Where does the "lost" energy go? Is it really lost?

Is friction ever useful? What about car brakes?

Measure the velocity of a rifle bullet by means of the ballistic pendulum.

Trace what happens to the energy of the bullet as it leaves the gun barrel.

Why does not the K.E. of the block = the K.E. of the bullet?

At what rate is the human machine capable of doing work, say when making a standing jump? i.e., can we get some idea what its maximum _power_ or _rate of doing work_ is, in ft.lb/sec, or gm.cm/sec?

FOURTH WEEK

FOR STUDY AND READING

ESSSENTIAL
    Lemon, pp. 57-64, 65-74, 83-90.
    Or Loeb and Adams, pp. 55-61, 109-120, 134-138, 178-187.

SUUGGESTED
    Knowlton, pp. 12-21, 26-33, 36-43.
    Saunders, pp. 79-86.
    Webster, Farwell, and Drew, pp. 70-82, 110-112.

Fifth Week

SECTION III - ENERGY (Continued)

C. **PRINCIPLE OF CONSERVATION OF ENERGY**

Just what does it really mean?
Does the total amount of energy involved in physical and chemical processes remain constant?

MECHANICAL ENERGY
    How does the principle apply to cases where mechanical energy is transferred from one form to another, or from one machine to another, as by a chain or belt drive, or through gears?
    Does the expenditure of a certain amount of mechanical energy necessarily result in the appearance of exactly that same amount of mechanical energy elsewhere?
    Conversion of mechanical energy into other forms.
        Frictional "losses." Mechanical energy converted to heat energy.
        When we speak of a machine having an efficiency of 50% do we imply that 50% of the energy has gone out of existence?
        Can we say in general that:
            Mechanical energy input = Mechanical energy output + amount converted to other forms of energy.

HEAT ENERGY
    Why has it been customary to measure heat energy in units different from those used for mechanical energy?
    Can we isolate a quantity of heat energy?
    The distinction between <u>quantity of heat</u> and <u>temperature</u>.

    <u>Temperature</u>.
        Do we have any basis for considering the <u>temperature</u> of a substance as fundamentally associated with the average kinetic energy of the molecules of the material? We shall study this question further in a later section.
        What types of physical phenomena are suitable for measuring temperature? Is the expansion of a substance like mercury necessarily the most accurate basis for thermometry?
        How do the centigrade and Fahrenheit scales differ?
            Why use the "<u>ice</u>" and "<u>steam</u>" points as the "<u>fixed</u>" points for a temperature scale?
            How can we convert temperature readings from centigrade to Fahrenheit, or vice versa?
        Why is the absolute, or Kelvin, temperature scale so important?

    <u>Heat Energy Units</u>.
        How do we use the <u>calorie</u>, or <u>British Thermal Unit</u> (BTU), in the measurement of heat energy?
        Is it logical to use the amount of heat necessary to raise 1 gm of H_2O, $1°C$ as a unit?
            What is the corresponding unit in the English system?

FIFTH WEEK

C. CONTINUED

Calorimetry.
Application of Conservation of Energy to heat transfers, involving only two systems.
    Loss of heat by one system = gain in heat of other system.
Basic law of heat change, where there is no change of state.
    Change in heat energy = (Specific heat)x(Wt)x(temp.change)
                     = $SW \Delta T$

Mechanical Equivalent of Heat.
Why were the experiments of Rumford and Joule which disproved the caloric theory of heat and proved that a definite amount of mechanical energy was equivalent to a definite amount of heat energy so important?
    Mechanical equivalent: 42,700 gm.cm = 1 calorie or
                                778 ft.lb = 1 BTU
What processes do we know which involve or depend on transfers of mechanical energy to heat energy, or vice versa?
    What about heat engines?

CHEMICAL ENERGY
Do chemical reactions always involve changes in energy?
    What happens to the energy (usually heat) absorbed or liberated in chemical reactions?
The chemist uses the terms <u>exothermic</u> and <u>endothermic</u> to designate ractions which liberate or absorb energy.
How is Conservation of Energy operating in the case of conversion of energy liberated in combustion of coal, gas, or gasoline into other forms of energy?
    Do we utilize the energy of other chemical reactions on a large scale, either directly or by transfer to some other form?
These questions will be considered much further next semester in Science A2.

WAVE ENERGY
What characteristics have the various types of wave energy, such as <u>water waves</u>, <u>sound waves</u>, and <u>light waves</u> in common?
    Do all waves require a physical medium for the transmission of energy?
    Considering sound energy as the transmission of mechanical energy impulses from molecule to molecule, why should one expect sound to travel slower (1100 ft/sec, 330 meters/sec) than light energy (186,000 miles/sec or 3×10^{10} cm/sec)?
    Why do we consider sound waves as <u>longitudinal</u> and light waves <u>transverse</u>?

In what ways are the terms:
    Frequency - no. complete vibrations or cycles per sec
    Period   - time for one complete vibration or cycle
    Amplitude - displacement from equilibrium position
    Wavelength - distance between points in same phase on two successive waves
    used to describe wave motion?

F I F T H   W E E K

C. CONTINUED

If the fundamental relation governing all wave motion is that:
velocity of propagation = frequency x wave length
what is the wave length of the radio wave from WEAF if the frequency is 660 kilocycles?
Under what conditions do we find that the <u>intensity</u> of wave energy (i.e., the energy falling normally on 1 cm^2 of surface) is inversely proportional to the square of the distance from the source? $I \sim \frac{1}{d^2}$?

Are the ears and eyes good instruments with which to measure sound and light energy? Where are they most in error, in estimating intensity or frequency?
Wave Energy Conversion.
What devices transfer wave energy to other forms or vice versa?
Consider the chain of energy conversions involved in the detection of the light energy from the projection lamp by a thermopile, amplifier, and galvanometer.

IN THE LABORATORY

How can we measure a quantity of heat energy? SW ΔT

Mechanical Equivalent of Heat.
Using the apparatus supplied, study the conversion of mechanical energy into heat energy. Determine the mechanical equivalent of heat, through the measurement of the work done in gm.cm (force x distance), and the heat energy transferred, as measured by the temperature rise of water and the known weight of metal parts of the calorimeter.
What factors account for the difference between the results obtained, and the accepted values for this relation between gm.cm and calories?

Energy Content of Illuminating Gas.
Study the measurement of the heat energy content of the illuminating gas supplied by the gas company, with the Junker type calorimeter. Note that it makes use of the same principles as the mechanical equivalent of heat apparatus, measurement of the temperature rise of a known quantity of water, caused by a known amount of gas. The fuel value is usually given in BTU per ft^3.

Efficiency of a Gas Burner.
Assuming that a good gas burner delivers heat energy at a constant rate, measure the rate at which the burner heats up a measured supply of water. By measuring the amount of gas used, and knowing the fuel value of the gas, determine the efficiency of a number of types of gas burner.

Temperature Measurement.
Study the measurement of temperature with a thermocouple. Calibrate it using the steam, ice, and liquid air points (-180°C). Use it to measure the temperature of solid CO_2, (dry ice). Assuming that the temperature at which it just glows red is about 500°C, by extrapola-

FIFTH WEEK

tion, observe the approximate temperature in parts of the gas flame. Would all thermocouples give the same calibration? What if two wires of the same kind were used?

FOR STUDY AND READING

ESSENTIAL
Lemon, pp. 74-83, 116-119, 135-141, 333-346.
Or Loeb and Adams, pp. 170-178, 207-217, 223-228, 230-235.
Or Knowlton, pp. 107-118, 245-256, 265-274, 320-338, 369-378.

SUGGESTED
Chase, "Men and Machines."
Kaempffert, "History of American Inventions."
Knowlton, on Utilization of Energy, pp. 230-238, 257-265, 275-288.
Lemon, pp. 372-384.
Lenard, "Great Men of Science" on Helmholtz.
Loeb and Adams, pp. 235-245.
Miller, "Science of Musical Sounds."
Richtmeyer, pp. 53-56 on Rumford and Joule.

Sixth Week

SECTION III - ENERGY (Continued)

C. **CONSERVATION OF ENERGY** (Continued)

ELECTRICAL ENERGY
Why not measure electrical energy in the same way we measure mechanical energy?

Measurement of Electrical Energy.
The Three Fundamental Concepts.
Potential difference.
Supplies the electrical field which drives the negative electrons along the conductor. Usually measured in volts - V.
Resistance.
A term used to cover the opposition to the flow of electrons along the conductor, analogous to friction. Usually measured in ohms - R.
Current.
The rate of flow of the electrons along the conductor - determined by the applied potential difference (volts), and the resistance (ohms). Determined by n_e, the net number of electrons which pass a given cross-section per second, multiplied by the charge on an electron. Usually measured in amperes - I.

Ohm's Law. $R = \frac{V}{I}$ A constant for a given conductor.
Is it reasonable to expect that the ratio of the voltage to current should be constant over a wide range of values?

Ammeters and Voltmeters.
The force on a conductor carrying current in a magnetic field is proportional to the current and at right angles both to the direction of the field and to the direction of the current.
How do we use this force to measure current in a simple galvanometer?
Why is the usual practice to provide ammeters with a low resistance in shunt, and voltmeters with a high resistance in series?

Electrical Power.
The watt. $VI = V^2/R = I^2R$.
The rate of utilizing energy.
One watt is equivalent to about .733 ft.lb. per second, or about 10,200 gm.cm per second.

Electrical Energy.
The product of p.d., current and time interval VIt, I^2Rt, or $(V^2/R)t$.
Watt-seconds, or more generally, kilowatt-hours.

SIXTH WEEK

C. CONTINUED

Equivalent to ft.lbs., or gm.cm.
   Which units appear on your monthly bill from the electric company?

Direct and Alternating Current, <u>DC</u> and <u>AC</u>.
  DC: constant voltage and current.
  AC: continually reversing voltage and current, sine wave form.
    What is meant by 60 cycle AC?
    Why must the peak value of 120 volt AC be larger than 120 volts, i.e., $\sqrt{2}\,(120)$ or about 170 volts?

Transformations of Electrical Energy.
  The "electrical age."
    The conversions of electrical energy to other forms are in general highly efficient, and are used most widely of all types of energy conversions.
      1 kilowatt-hour = about 2,640,000 ft.lb
                     = about 900,000 calories
  Transformations into <u>heat</u> and <u>light</u>
    The heat developed in a resistance = $.24\ I^2Rt$ calories.
    How much of the energy supplied to a hot lamp filament is converted to visible light?
  Batteries - sources of EMF.
    Dry cells - storage batteries.
<u>The generator.</u>
  Transfer of mechanical to electrical energy.
  Electrical and magnetic fields.
    The magnetic field produced by the current flowing through a wire.
    The electromagnet, increased field strength due to the iron core.
  Induced E.M.F.'s. Faraday's discovery in 1830 of the voltage induced in a wire moved through a magnetic field.
  The essential parts of a generator.
    The magnetic field: Field coils - iron frame.
    The armature: Coils of wire rotating in the magnetic field, in which voltage is induced.
    Commutator and brushes: Reversing connections to armature coils at the proper moment so as to provide unidirectional flow of electrons in the external circuit of a direct current (DC) generator.
    Or slip rings and brushes: Continuous connection to armature coils so as to produce alternating voltage; AC generator.
    And of course, a source of mechanical energy, steam, water power, etc. to rotate the armature.
  What determines how much mechanical energy must be supplied to rotate the armature?
  Is electrical energy "produced"?
<u>The motor</u>.
  Transfer of electrical to mechanical energy.

SIXTH WEEK

C. CONTINUED

        Based on the companion effect to that of the generator, i.e., the force exerted on a wire carrying current in a magnetic field.
           How does the motor differ from the galvanometer?
        The essential parts of a DC motor.
           The magnetic field: Field coils - iron frame.
           The armature: Coils free to rotate in the magnetic field due to the force exerted on them when current flows through them.
           Commutator and brushes: To reverse the current through the armature at the proper moment so as to provide continuous rotation in the same direction.
           And of course, a source of electrical energy.
        Counter E.M.F.
        Why can a DC motor serve as a generator if its armature is rotated?

    Distribution of Electrical Energy.
        Long-distance transmission: The necessity for high voltage, low current transmission to reduce I^2R power losses in the resistance of the lines.
        The alternating current step-up and step-down transformer.
           Makes possible transmission of high voltage, low current over long distances, and stepping down to low voltage, high current for use in local distribution systems.
        Why can't we use transformers on direct current?

    Utilization of Electrical Energy.
        Energy conversions: Follow the energy changes from the sun's energy, to prehistoric plants of the Carboniferous era, to coal under a boiler, to the distribution of electrical energy to light cities, drive trains, and run industrial establishments.
        Large-scale energy projects.
           Why do these huge plans, such as the TVA development, or the Boulder Dam project, always involve electrical energy?
        Is "cheap power" a possibility?

    Why does electrical energy appear to be so fundamental?
        Is the distinction between electrical energy and the other forms of energy real?
        Does the electrical nature of matter foreshadow future developments?
        Is it possible to "store up" energy? On a small scale; on a large scale? Or for a short time; for a long time?

IN THE LABORATORY

Examine the various forms of galvanometers, ammeters, and voltmeters, to learn as much as possible about the purpose and principle involved.

SIXTH WEEK

Note the contrast between the old instruments of the moving-magnet, fixed-coil type, and the modern instruments of the moving-coil, fixed-magnet, or D'Arsonval type.
On what factors does the sensitivity of the suspension type galvanometer depend? What sets a limit to the obtainable sensitivity?
Study the various ammeters and voltmeters, and draw typical diagrams of the two types, with appropriate resistances in series or parallel.

With suitable ammeters and voltmeters, measure the potential difference (volts) between the terminals of various types of electric lamps, and the corresponding currents.
Measure simultaneously, by means of an illuminometer of the photometer or photo-cell type, the illumination in candle power of the various types of lamps.
Calculate the resistance of the lamp filaments when hot, and the energy in watt-seconds or K.W.H. supplied per hour of service.
Compare the illumination produced per watt by the various types of lamps (i.e., the relative efficiency).
How is the high efficiency of the "photo-flood" lamps obtained?

Measure the electrical energy, VIt (watt-seconds), supplied to a direct current motor, and at the same time measure the mechanical energy, Fd (gm.cm), turned out by the motor, at various loads.
If one watt-second is equivalent to about 10,200 gm.cm, how does the motor efficiency vary with the loads used?
What happened to the "lost" energy?

Using an electrical calorimeter, measure the quantity of heat energy in calories developed by a definite amount of electrical energy (watt-seconds). What is the relation between the units used to measure the two forms of energy?

FOR STUDY AND READING

ESSENTIAL
    Lemon, pp. 222-262, 265-268.
    Or Knowlton, pp. 320-339, 369-379, 379-390.
    Or Loeb and Adams, pp. 324-328, 333, 373-379, 383-390.

SUGGESTED
    Hodgins and Magoun, "Behemoth."
    Knowlton, on Utilization of Energy, pp. 230-239.
    Loeb and Adams, on Heat Death, pp. 235-245.
    Pupin, "From Immigrant to Inventor."
    Richtmeyer, on Faraday, pp. 60-70.
    Saunders, on Generators and Motors, pp. 400-410.

Seventh Week

SECTION III - ENERGY (Continued)

C. **CONSERVATION OF ENERGY** (Continued)

    Conservation of Energy and Transformations of Energy.
        How has man's utilization of energy progressed since the beginning of time?
            Why has the increase in the use of energy gone further since 1900 than in all the preceding centuries?
            What further effects on social and political systems can we expect if the utilization of energy continues to increase at its present rate?
        Viewed in the light of your study of various phenomena as outlined in previous sections, how fundamental to modern civilization are these facts about energy?
            Do you expect energy to become more important or less important in the future than it is at present?
            Can any appreciable fraction of our human activities be independent of these phenomena based on or utilizing energy?
    Broader aspects of the conservation principle.
        Is it basic in all phenomena, or does it break down?
            Are we free from the perpetual motion idea?
        The First and Second Laws of Thermodynamics.
        Availability of Energy.
            How can we have energy, and yet not have it available?
            Is the available energy of our earth decreasing continually?
                How about our coal, gas, oil, etc.? Will we eventually use up our supply?
            Is the universe as a whole running down?
            Will we eventually suffer "heat death"?

D. **THE PARTICLE NATURE OF MATTER**

    With the background gained from our studies of energy, let us look back over our early considerations of the nature of matter and enlarge our understanding of the atom, as a basis for the kinetic theory of matter, and the relation between matter and energy.

    Early Speculation.
        What did Democritus and other early thinkers really know?
        How did the first experimental evidence grow out of the work of Dalton, Boyle, and Avogadro on the nature of gases?

    Further Development of Atomic Hypotheses.
        Atomic Structure.
            The hard spherical atom.
                Why did the concept of the atom as an indivisible, hard, spherical, elementary particle of the element persist so long?
                Was this concept sufficient for most of the early work on gases, liquids, solids, and chemical reactions?

SEVENTH WEEK

D. <u>CONTINUED</u>

    Ideas about internal structure.
        How did the idea of a definite internal structure for the various atoms begin to grow out of the early experiments of Geissler, Hittorf, and Crookes on electrical discharge in gases, culminating in the discovery of the electron by Sir J. J. Thomson in 1897?
    Atom models.
        Why does the scientist attempt to build a model?
        When should we be satisfied with a model?
            To what fraction of us do our models represent absolute reality?
    Have atomic hypotheses helped or hindered discoveries?
    Early atom models.
        The Thompson atom. Positive charge distributed throughout the atom. Displaced by the Rutherford nuclear atom.
        The Lewis-Langmuir atom. A static geometrical structure, useful for some chemical interpretations but now largely discarded.
    Rutherford-Bohr nuclear atom model.
        The nucleus.
            How did the experiments of Rutherford, firing alpha particles (helium nuclei) at various atoms, establish the idea of a tiny, positively charged central nucleus, containing most of the mass of the atom?
            The internal structure of this complex "central sun," having a range of radius of about $2\text{-}8 \times 10^{-13}$ cm from hydrogen to uranium, now appears to consist of protons (hydrogen nuclei, approximately 1.007 mass units, one + charge), and neutrons (approximately same mass as proton, but no electrical charge).
        The external "planetary" electrons.
            Mass about 1/1840 that of proton, one - charge.
            Equal in number to the protons in the nucleus, since the atom is electrically neutral.
            Early ideas on electron "orbits," electron "shells."
        A dynamic atom.
            Necessary to produce a stable atom, with the negative electrons in equilibrium under the electrical attraction inward to the positive nucleus, and the centrifugal force outward.
    These ideas, together with the modifications of Bohr, and others we shall consider further in the section on <u>Radiation</u>.

<u>Periodic Relations among the Elements</u>.
    Complex Atoms.
        Development from the simple $_1H^1$ atom with one proton and one electron, to the more complex atoms up to the uranium atom, $_{92}U^{238}$, with a nucleus of 92 protons, and 146 neutrons, and 92 external "planetary" electrons grouped in several "shells."

SEVENTH WEEK

D. CONTINUED

Atomic Weight.
> The weight of the atom in mass units based on O^{16} as 16.0000.
> Why does the atomic weight depend almost entirely on the nucleus?
> Atomic weights of the various atoms are nearly whole numbers.

Atomic Number.
> The charge on the nucleus, i.e., the number of nuclear protons, or external electrons.

Isotopes.
> The early experiments.
> The work of Aston, Dempster, Urey, and others.
> Practically all elements are families of atoms, having:
>> Nearly the same physical and chemical properties.
>> The same atomic number, i.e., same number of protons in nucleus, and same number of external electrons, and almost identical extranuclear structure;

But:
> Contain different numbers of neutrons in the nucleus, therefore have different atomic weights.
>> For example, we have:
>> $_{82}Pb^{204}$, $_{82}Pb^{205}$, $_{82}Pb^{206}$, $_{82}Pb^{207}$, $_{82}Pb^{208}$, $_{82}Pb^{209}$
>> and also other isotopes in very small quantities.
>> Observe the notation which indicates both atomic number and atomic weight.
> Hydrogen and deuterium - Urey
> $_1H^1$ and $_1D^2$
> Heavy water D_2O

Chemical Atomic Weights.
> The average of the weights of the mixture of isotopes or a given element.
>> Ordinary lead averages about 207.22 for example.

Mendelejeff's Periodic Table. Are there others?
> A large number of physical and chemical properties, such as density, atomic spectra, boiling point, chemical valence, etc. vary periodically among the elements.
> Have our ideas of the periodic nature of the elements progressed since Mendelejeff?

E. THE KINETIC THEORY OF MATTER

How certain is our experimental basis for the belief that the molecules of all matter are in rapid and continual motion, and that temperature is simply a measure of the average kinetic energy, $1/2(W/g)v^2$, of the molecules?

How can we measure the effect of a large number of molecules if we cannot know what is happening to every one of them at a particular time?

Why are our experimental and theoretical studies of matter in general only statistical?

SEVENTH WEEK

E. CONTINUED
Gases.
Characterized by Very Small Interatomic Forces.
The atoms make frequent elastic impacts with each other, much like elastic balls, but are in contact with their neighbors only a small fraction of the total time.

Brownian Movement.
How does the continual jerky motion of microscopic pollen grains, first observed by Brown, show the behavior of the molecules buffeting them about?
Do we see the molecules themselves?

Development of Ideas About Gas Pressure.
Pressure = Force/area = force per unit area (in gm/cm^2, or lb/in^2, etc.)
How did Joule's work on the Mechanical Equivalent of Heat lead to his idea that gas pressure was due to the constant bombardment of the walls of a container by the gas molecules?
Can we prove experimentally that the relation between gas pressure and the average kinetic energy of the molecules is such that
$P = \frac{1}{3} \frac{wv^2}{g}$, where w = weight/cm³ or density.

The velocity of the average air molecule (N_2 or O_2) at ordinary temperatures is about 1600 ft/sec.

All types of molecules of whatever weight have the same average kinetic energy $1/2(w/g)v^2$, at the same temperature.
How much faster does a hydrogen molecule H_2 travel than an oxygen molecule O_2, in order that the kinetic energy shall be the same?

IN THE LABORATORY

Measurement of Pressure.
Investigate the principle involved in the operation of the various forms of pressure measuring devices available, such as mercury manometers, Bourdon gauges, mercury and aneroid types of barometers, airplane altimeters, airplane speed indicators, etc.

Measurement of Barometric Pressure.
Why does the barometer measure the weight of the air above it?
Read the barometer in the laboratory, on the 1st and top floors.
Are the readings consistent?
Can we compute the vertical distance between the barometers?
What do you know about barometric pressure, as used in weather maps, and weather forecasting?

Brownian Movement.
Study, under proper illumination, with a high power microscope, the motion of
 1 Smoke particles in air.
 2 Gold particles in water.
What types of path do the particles buffeted about by the high speed molecules travel?

SEVENTH WEEK

How good is this as a foundation for the kinetic theory?
Would you expect the wing of an airplane to show a large Brownian movement?

Boyle's Law.
If sufficient time is available, determine how accurately Boyle's law, $P_1V_1 = P_2V_2$, enables you to compute the volume of air in a tank under pressure.
Observe by pumping air into a steel tank of known volume to a high pressure, and then measuring the volume of air at atmospheric pressure by allowing the air to escape through a gas meter.

FOR STUDY AND READING

ESSENTIAL
Bazzoni, "Energy and Matter," pp. 18-48.
And Loeb and Adams, pp. 219-223, 260-267, 283-286.
Or Lemon, pp. 95-105, 149-157, 308-315, 325-330.

SUGGESTED
Bazzoni, "Energy and Matter," pp. 1-18.
Knowlton, pp. 143-155.
Lenard, "Great Men of Science" on Boyle, Watt, Dalton, Joule, Clausius, and Kelvin.
Loeb and Adams, pp. 272-274, 286-288, 290.

Eighth Week

SECTION III - ENERGY (Continued)

E. **THE KINETIC THEORY OF MATTER** (Continued)

<u>Gases</u> (Continued)

How have our ideas developed from the work of Dalton on the partial pressures exerted by mixtures of gases, the compression of gases by Boyle, the ideas of Avogadro, and on up to the refined theories of Clausius, Maxwell, and others, viewing gases from a statistical mechanical point of view?

Avogadro's Hypothesis.

All gases under the same conditions contain equal numbers of molecules.

Boyle's Law.

The product of pressure and volume, (if temp. is constant),

$$P_1V_1 = P_2V_2 = \text{a constant} = \frac{1}{3}\frac{W}{g}v^2$$

How well do gases follow this law?

Are there any assumptions involved in Boyle's law that are not quite correct?

<u>Temperature</u>.

The molecules of all gases have the same average kinetic energy, $1/2\ (w/g)\ v^2$, at the same temperatures.

The average velocity of an O_2 molecule in air is about 1600 ft/sec under ordinary conditions.

The H_2 molecule weighs 1/16 as much as the O_2 molecule.

The average velocity of an H_2 molecule is therefore $\sqrt{16}$ or 4 times that of the O_2 molecule, about 6400 ft/sec.

Absolute zero.

The pressure of an ideal gas approaches zero linearly at about -273.1°C as the temperature is lowered, volume being kept constant. The volume behaves in the same way, if pressure is kept constant. That is, molecular kinetic energy, $1/2(w/g)v^2$ approaches zero at that temperature.

Absolute, Kelvin or Thermodynamic scale of temperature.

Using -273.1°C as the zero point of a new scale, the pressure and volume of an ideal gas become linear functions of T.

If the volume is held constant:

$$P = \frac{1}{3}\frac{D\ v^2}{g} \text{ is proportional to T}$$

or $\frac{P_1}{P_2} = \frac{T_1}{T_2}$ Charles or Gay-Lussac's law

If pressure is held constant:

V is proportional to T

or $\frac{V_1}{V_2} = \frac{T_1}{T_2}$

EIGHTH WEEK

E. <u>CONTINUED</u>

General gas law.
Using these relations, and Boyle's law:

$$PV = RT = \frac{1}{3}\frac{W}{g}v^2 \qquad \begin{array}{l} R = \text{General gas constant} \\ = 84,800 \text{ cm/deg.C} \end{array}$$

or $\frac{P_1 V_1}{P_2 V_2} = \frac{T_1}{T_2}$ or $= \frac{W_1 v_1^2}{W_2 v_2^2}$ or $= \frac{W_1 T_1}{W_2 T_2}$

The last two expressions are useful when the weight of gas changes.
Note that <u>absolute</u> and <u>not</u> gauge pressure must be used, and that likewise <u>absolute</u> and <u>not</u> centigrade temperatures must be used.
How do we use these relations in actual problems?
What are the limitations of the general gas law?

Kinetic Theory Concepts.
Are all the molecules of a gas moving alike?
Mean Free Path.
The average distance a molecule travels before it makes a collision with another molecule.
Usually:

$$\text{M.F.P.} = \frac{1}{\sqrt{2}\, N\, \pi\sigma^2} \qquad \begin{array}{l} N = \text{no.mol/cm}^2 \\ \sigma = \text{diameter of a molecule} \end{array}$$

The N_2 (nitrogen) molecule has a M.F.P. of about 0.00001 cm under ordinary conditions in air.
The average velocity is around 1600. ft or 52,500 cm/sec in air, under ordinary conditions, 76 cm Hg pressure.
If the M.F.P. is therefore inversely proportional to the pressure, what is the M.F.P. of N_2 under high vacuum conditions, say around .0000001 mm Hg?

Maxwell distribution
What does a typical Maxwell distribution curve show us regarding the distribution of molecular velocities?
Do any have zero velocity?
Would the predictions of such a distribution curve be accurate if we had only 10 molecules to deal with?
Why must we have large numbers of individuals?
How is the so-called Gaussian error curve related to such considerations?
How are the ideas underlying the distribution of molecular velocities, and other features of molecular motion applicable to other fields?
How does the field of <u>statistics</u> and <u>statistical analysis</u>, now used so widely in many lines, such as the numerous government reports, life insurance figures, and the various other forms of industrial, financial, economic, and social studies, use the developments from the mathematical analyses growing out of these ideas?

EIGHTH WEEK

E. CONTINUED

Gas Thermometers.
Why are the characteristics of a nearly perfect gas such as hydrogen useful in thermometry?
Constant volume thermometer.
The volume of the gas inclosed in a bulb is kept constant by changing the pressure by means of a mercury column, so that
$$P_1/P_2 = T_1/T_2$$
determines absolute temperature of the inclosed gas if one temperature is known.
Constant pressure thermometer.
The pressure on a volume of inclosed gas is kept constant so that the change in volume measures T if one temperature is known, since
$$V_1/V_2 = T_1/T_2$$

Transport Phenomena.
Diffusion.
Transport of molecules through a gas.
If the average velocity of molecules of illuminating gas is more than 2000 ft/sec, why doesn't it diffuse instantly throughout a room?
How do the average molecular velocities and the mean free path between collisions control the rates of diffusion of gases?

Viscosity.
Determined by the transfer of momentum.

Conduction of heat, sound.
Determined by the transfer of energy from molecule to molecule.
Dewar flasks - vacuum thermos bottles.
Vacuum to reduce conduction by gas molecules.
Silvered to reduce radiation losses.

Dynamics of Gases.
Bernoulli effect.
A seeming paradox - the reduced pressure in gases moving at high speeds.
$$P_1 - P_2 = 1/3(D/g)v_1^2 - 1/3(D/g)v_2^2$$
Air resistance - streamlining.
Back pressure due to Bernoulli forces at high speeds.
"Skin friction" along a surface
Influence of viscosity.
How important is streamlining in practice?

High Vacua.
Why do we build apparatus and do experiments in vacuo?
Vacuum pumps.
Early mechanical or barometer type pumps.
Modern mechanical rotary oil pumps.

EIGHTH WEEK

E. CONTINUED

        High speed molecular diffusion pumps.
           Gaede-Langmuir.
           Diffusion of residual gas into stream of high velocity mercury or oil molecules.
      McLeod gauges - liquid air traps to condense residual vapors.

    Pressures.
        The pressure in a good X-ray tube may be as low as .0000001 mm of Hg. How many times does an O_2 molecule bounce back and forth between the walls before it hits another gas molecule?
        How many molecules are still present per cm^3, if there are about 27×10^{18} mol/cm^3 at 760 mm Hg.

Liquids.

    Characterized by intermolecular forces much larger than in gases, but not sufficient to produce a definite shape.
    Do liquids still show the same effects of pressure, volume, and temperature as gases?
        The compression of liquids even under tons per sq.in pressure is very small.
        Change of density with temperature relatively small.
        What about viscosity? Why use "light" oil in winter?
    Surface Tension.
    Vapor Pressure.
        The average kinetic energy of the molecules in the liquid state is still proportional to the absolute temperature.
        How does vapor pressure change with temperature?
      <u>Boiling point</u>: Liquids boil when the vapor pressure is equal to the pressure of the gas above it.
      <u>Saturated vapor pressure</u>: Equilibrium vapor pressure above an inclosed liquid is dependent only on the temperature.
      <u>Relative humidity</u>:
        A ratio: <u>Actual pressure of vapor present</u>
               Saturation vapor pressure
        Why does a lowering of temperature produce precipitation?
        What did the Weather Bureau mean by 105% relative humidity last summer?
    How well does the kinetic theory explain the phenomena?

Solids.

    Enormous intermolecular forces.
    Molecules elastically bound to approximately fixed positions.
    The molecules are free to vibrate about these fixed positions with average kinetic energy proportional to temperature, as before.
    Further studies of structure in solid state will be made next week.

How well does the kinetic theory picture satisfy you regarding the three states of matter?

EIGHTH WEEK   31

IN THE LABORATORY

Constant Volume Thermometer.
  Calibrate a constant volume thermometer using the usual "fixed
  points," ice 0°C (273° Abs.), and the steam point 100° (373° Abs.)
  at 760 mm Hg. barometric pressure. If volume is kept constant:
  $$P_1/P_2 = T_1/T_2$$
  Use the thermometer to measure room temperature, the temperature of
  a number of molten solids or liquids at high temperatures, such as
  sulfur, and also try it for low temperatures solid CO_2 (-80°C) and
  liquid air (-185°C).
  In each case, where possible, check the results with a mercury-in-
  glass thermometer.

Constant Pressure Thermometer.
  Use a constant pressure thermometer likewise to measure the change in
  volume of the inclosed gas at various temperatures, determining the
  temperature from
  $$V_1/V_2 = T_1/T_2$$
  What limitations do the gas thermometers have?
  Be sure you understand how to <u>use</u> the gas laws.

Electrical Methods for Temperature Measurement.
  Simultaneously, study the temperature measurement of the various ma-
  terials with a thermocouple, or a platinum resistance thermometer.
  What advantages do these electrical methods have?

The Molecular Beam Laboratories.
  Visit some of the laboratories where Professor Rabi's students are
  working on molecular beam problems.
  Note the technique of obtaining beams of molecules by vaporizing a
  substance in a small electrically heated oven and allowing the
  molecules of the vapor to issue from the oven through a narrow slit.
  Observe the vacuum systems, with the large diffusion pumps, mechanical oil pumps, water cooling, and liquid air traps used to provide
  the low pressures so that molecules may travel long distances with-
  out colliding with gas molecules.
  Note also the methods of studying the properties of molecules by
  their deflection in magnetic fields.
  What are some of the methods used to detect beams?

FOR STUDY AND READING

ESSENTIAL
  Lemon, pp. 112-134, 157-169.
  Loeb and Adams, pp. 267-270, 272-282.

SUGGESTED
  Bragg, "Concerning the Nature of Things," p. 46.
  Jauncey, "Modern Physics," pp. 149-167, 167-176.
  Knowlton, "Kinetic Theory," pp. 155-166; "Physics of Air," pp. 175-
    189.
  Lenard, "Great Men of Science," on Clausius, Kelvin, and Maxwell.
  Webster Farwell and Drew, pp. 213-219, 241-246.

Ninth Week

SECTION III - ENERGY (Continued)

E. **THE KINETIC THEORY OF MATTER** (Continued)

Continuity of State.
　The transitions of substances between the gaseous, liquid and solid states.
　　What is the significance of a change of state?

Heat of Fusion.
　The energy required to change 1 gm of solid into the liquid state, the work being done against those forces which hold atoms together in fairly definite configurations.
　　Water:
　　　About 80 calories must be added to convert 1 gm of ice at $0°C$ to 1 gm of water at $0°C$.
　　　Conversely, 80 calories must be removed from 1 gm of water at $0°C$ to convert it to 1 gm of ice at $0°C$.

Heat of Vaporization.
　The energy required to change 1 gm of liquid into the vapor state, the work being done against internal forces which exist in the relatively dense liquid and against external forces, e.g., force due to atmospheric pressure.
　　Water:
　　　About 540 calories must be added to convert 1 gm of water at $100°C$ to 1 gm of steam at $100°C$.
　　　Conversely, 1 gm of steam at $100°C$ condensing into 1 gm of water at $100°C$ liberates 540 calories.
　　Note that, in general, a substance to which heat is being added remains at a constant temperature at the fusion or vaporization point until all of the substance has been converted.

Heat of Sublimation.
　The energy required to liberate molecules from the solid state directly to the gas state.

Why should pressure have an effect on a change of state?

Liquefaction of Gases.
　In general, a combination of low temperatures and high pressures is required to liquefy gases.
　The classical experiments of Andrews on CO_2.
　　Ordinary Boyle's law experiments may liquefy CO_2 if done below $31°C$, the critical temperature.
　Critical temperature.
　　The temperature above which the application of higher pressures will not liquefy the gas.
　　　O_2 $-118°C$, N_2 $-146°C$, H_2 $-241°C$
　Production of liquid air.
　　O_2 $-193°C$, N_2 $-195°C$. The Linde process.
　Liquid hydrogen and helium.
　　$-252.8°C$ or $20.3°K$; and $-268.8°C$ or $3.3°K$.

NINTH WEEK

E. CONTINUED

How near shall we approach absolute zero?
Mechanical refrigeration.
Cooling due to energy absorbed (heat of vaporization) by substance in liquid state upon expansion to the vapor state.

Thermal Cycles.
Cycles in thermal processes.
The Carnot Cycle.
A reversible cycle.

Heat Engines.
Typical thermal cycles in heat engines, such as steam engines, turbines, or gasoline engines, - all based on conversion of heat energy to mechanical energy.
Gas at high pressure and temperature into the cylinder.
Gas does work against piston, expanding to lower pressure and temperature.
Exhausted at low pressure and temperature.
Condensed and returned to boiler, in steam engine case.

$$\text{Maximum efficiency} = \frac{T_1 - T_2}{T_1}$$

$$= \frac{\text{Boiler temperature} - \text{Condenser temperature}}{\text{Boiler temperature}}$$

Temperatures here are on Kelvin scale.
Approaches 100% only when temperature of exhaust gas or steam is at absolute zero.
45% most feasible under ordinary conditions, with "perfect" engine.

Refrigeration Machine Cycles.
Is a refrigeration machine a heat engine run backwards?
The essential parts of a refrigeration machine.
Motor-compressor to condense gas to liquid.
Heat radiator to radiate heat due to compression of gas.
Expansion chamber, cooled region due to gas expansion.
System to reintroduce gas to compressor.

Available Energy.
The unavailability of heat energy in general, after having passed through a heat engine, unless it has a temperature higher than the surroundings, even though it is far above absolute zero.
The Second Law of Thermodynamics.
It is impossible to transfer heat from a cold to a hot body by a self-acting mechanism.
All earthly processes tend to go in such a direction as to degrade heat energy.
Are we approaching a dead level of temperature as we use up our available energy?
Can we say definitely that the universe is running down?

NINTH WEEK

E. CONTINUED

    Structure in Matter.
        Review previous evidence.
        Crystal forms and changes in crystal forms.
            Non-crystalline forms?
                Is there also structure in liquids? In gases?
                Apparently some structure exists in all forms of matter.
        Solid state.
             Cubic, hexagonal, tetragonal, rhombic, orthorhombic, etc.
            Observe the various crystal models in the case on 8th floor of the Pupin Physics Laboratories.
            What methods do we have for studying crystal structure?
            Crystal structure and physical properties.
                The difference in structure of a crystal in different directions also produces changed physical properties in the various directions.
                Mechanical differences.
                    Density, strength, hardness, elasticity, ductility, etc. all depend greatly on crystal form and size.
                  Metallurgists must study this point closely.
                Optical properties.
                    Index of refraction, polarization, depend on structure.
                Electrical conductivity, magnetic susceptibility, permeability, saturation, etc., thermal expansion coefficients, the speed of sound; all such factors are closely related to crystal structure.
                X-ray scattering or reflection is particularly useful to study crystals, as discussed in a later section on Radiation.
            The production of materials with the desired properties depends greatly on knowledge of molecular structure.
            Our knowledge of the solid state is still far from satisfactory.

    Implications of the Kinetic Theory.
        Statistical nature of kinetic theory laws.
            Can we say that in general all other physical laws are likewise statistical, since they involve the behavior of a group of individual units?
            What can be expected of a statistical theory?
                To what extent can such a theory be trusted?
        Macro-physics and **Micro-physics**
            Do these have parallels in other fields, such as economics, sociology, medicine, or biology?
            What do we lose by generalization?
            The dangers involved in extrapolation, in passing from the statistical result to the individual case, or vice-versa.

                      IN THE LABORATORY

Change of State.
    Ice to water. Using a mercury-in-glass thermometer, make a tempera-

NINTH WEEK

ture - time record as the ice-water mixture is heated at as constant a rate as possible. Stir the mixture constantly to maintain a uniform temperature throughout, and continue taking readings until after the ice has been melted for some time.
> Why does the temperature stay constant at 0°C until all the ice has melted?
> If the pressure in the room should be doubled or halved, what changes would occur in your record?

Melting tin or lead. Make the same type of time-temperature record for tin or lead heated in a graphite crucible. Use a thermocouple for the temperature measurement.

Vapor Pressure of Water.

Since boiling occurs when the vapor pressure of the substance is equal to the pressure of the gas above it, the apparatus supplied can be used to measure the vapor pressure of water at various temperatures. The pressure is lowered by means of a water aspirator and measured by a mercury manometer. The temperature of the boiling liquid is read on the mercury thermometer.
> Plot a vapor-pressure temperature curve after obtaining as many points as feasible in the time available.
> Compare the results with standard data. (W, F & D. pp. 261.)
> Does the vapor pressure of water ever become zero?

Refrigeration Machines.

If available, study the mechanism of a mechanical refrigerator.
> Observe the compressor, the radiating system for radiating the heat from the compressed gas, the valve to permit expansion of gas into the system of cooling coils, and the return system back to the compressor.
> Be sure to understand the thermal cycle involved.

Engineering School Laboratories and Power Plant.

Visit the mechanical engineering laboratories to observe the construction of steam boilers; reciprocating steam engines, their flywheel, connecting rod, piston assembly, the valve and condenser system; steam turbines, etc.

In the power plant, observe the boiler systems, automatic stokers, the main steam lines, the large reciprocating engines driving the generators, the switch boards, the Diesel motor-driven generators.
> Be sure you understand the thermal cycles involved.

FOR STUDY AND READING

ESSENTIAL
Lemon, pp. 141-148, 164-169, 169-173.
Webster Farwell and Drew, pp. 241-264, on Change of State.
Kimball, pp. 317-324, on Liquefaction of Gases.
Or Loeb and Adams, pp. 301-311.

NINTH WEEK

SUGGESTED
- Humphreys, "Physics of the Air."
- Jaffe, "Physics of Crystals."
- Kimball, pp. 324-328, on Heat Engines.
- Knowlton, "Energy of Winds and Water," pp. 230-238. "Heat Engines," pp. 166-174, 190-214.
- Mayer, "Seven Seals of Science," p. 235.
- Stuart Chase, "Men and Machines."

Tenth Week

SECTION IV - RADIATION

A. THE BEGINNING OF MODERN PHYSICS

The progress of science in the period before 1900.
    Why should the opinion have been held by some that all the major discoveries in physics had been made?
The opening of a new era.
    Why did such discoveries as the <u>electron</u>, <u>radioactivity</u>, x-rays, and electromagnetic waves, in this period open up so many new fields?
    What contributions did such experimenters as Thomson, Becquerel, Roentgen, the Curies, Hertz, Marconi, and others make to this development?
In this section we shall consider the unfolding of this new era.

B. THE ELECTRON

<u>The Discovery of the Electron - 1897.</u>
    Early experiments on electrical discharge in gases at low pressure.
        Geissler tubes, Hittorf, Crookes, and others.
    Cathode Rays.
        "Rays" coming from the cathode (- electrode) to the anode (+ electrode) in a discharge tube.
    The wave-particle controversy.
        Were they waves, or particles?
    Experiments of Sir J. J. Thomson.
        Deflection of the cathode rays by electrostatic field between two electrically charged plates, in a circular path, toward the + plate, i.e., the particles had a - charge.
        Deflection in a magnetic field.
            Deflected at right angles to the magnetic field, and to its original direction of motion, in a circular path.
    Measurement of ratio of charge to mass, e/m, of the cathode ray particles or <u>electrons</u>, by the combined effects of electrical and magnetic fields at right angles to each other.
    Electron:
        e/m = 1850 times that of the hydrogen nucleus, i.e., if the charge e is the same, the mass m is 1/1850 that of the H nucleus or proton.
    Charge on the Electron.
        How did Millikan measure the charge on the electron, using the effect of the electrical field between two charged plates on the rate of fall of charged oil droplets?
    What factors enter into the determination of this charge?
    The charge on the electron,
        e = about 4.77×10^{-10} ESU (electrostatic units)
            about 1.59×10^{-19} coulombs.
    The electron the elementary unit of negative charge, a unit quantity of electricity.
    Electron mass about 9×10^{-28} gm.

TENTH WEEK

B. CONTINUED

<u>Conduction of Electricity in Gases</u>.
    Is the current through a gas essentially different from that in a copper wire?
    Ionization.
        Ions: Charged particles.
            + ions, atoms with one or more electrons stripped off.
            - ions, electrons (or neutral atoms with extra electron attached momentarily).
                Recombination of ions to form neutral atoms usually takes place quickly.
        Ionization by impact.
            Ionization by electron impact occurs if electron is attracted by a strong enough electrical field, so that it gains enough energy to ionize a gas atom before it strikes the atom, i.e., it must gain sufficient energy to ionize in an average mean free path between collisions.
            Electrical discharges build up if ionization by impact occurs faster than recombination of ions.
            The initial free electron to produce ionization by impact probably comes from cosmic radiation or local radioactivity.
        Ionization also is produced by high energy protons, alpha particles (helium nuclei), beta particles (electrons), x-rays, gamma rays, etc.
    Electrical Discharges.
        Breakdown in air.
            Under ordinary conditions, an electrical field intensity of about 30,000 volts per cm is necessary to produce ionization by impact.
        At low pressures.
            If pressure is lowered, mean free path of electrons is increased, and less electrical field intensity is required to produce ionization.
        Discharge tube phenomena.
            The cathode glow, Faraday dark space, striations, the positive column.
    The positive "rays" in discharge tubes, and the characteristic radiations emitted by ionized or excited atoms will be considered later.

<u>Electrical Circuits</u>.
    Conduction in liquids. Ionization in solutions will be considered in Science A2.

<u>Conduction in Solids</u>. See previous discussion.
    Metallic Conduction.
        Between tightly packed atoms whose outer electrons overlap into neighboring atoms, a progressive interchange of electrons occurs if an electrical field is applied.
    Insulators.
        A progressive interchange of electrons between adjacent atoms occurs with very great difficulty if at all.

T E N T H W E E K   39

B. CONTINUED

Review of Elementary Concepts of Circuits.
Flow of electrons through circuits under **steady** conditions.
Ohm's law - $R = V/I$. Amperes, volts, ohms.
Resistances in series.
  Current I same throughout circuit - $I = V/R$
  Total resistance $R = R_1 + R_2 + R_3$ etc.
  The voltage across the various resistances, $V_1 = IR_1$; $V_2 = IR_2$
Resistances in parallel.
  The effective resistance is given by
    $1/R = 1/R_1 + 1/R_2 + 1/R_3$ etc.
  The voltage across each resistance is the same.
  The total current is given by $I = V/R$
  The current through each resistor is inversely proportional
    to the resistance, $I_1 = V/R_1$ etc.
Kirchoff's second law.
  The **algebraic** sum of all the intrinsic EMF's (volts) and the
    voltage drops across resistances, I_1R_1, I_2R_2, etc. around
    any closed path, is equal to zero.
       $\Sigma V = \Sigma IR$

Inductance - Electron Flow under **Unsteady** Conditions.
Electrical inertia. Analogous to mass, W/g, in mechanical systems.
Changing currents induce counter EMF's which oppose the current
  change.
    Counter EMF $V = L \Delta I/\Delta T$ (L = inductance)
      $L = \underline{1\ Henry}$ if current change of 1 ampere per sec. in-
        duces a counter EMF of 1 volt.
Inductance primarily associated with coils, but actually all wires
  have inductance.
Iron cores greatly increase a coil's inductance.
In inductive circuits, when the source of EMF is connected, the
  current gradually (exponentially) approaches its final value I,
  as given by $V = IR$.
    Conversely, when the switch is opened, the current gradually
      (exponentially) decays to zero.
Why is <u>Joseph Henry</u> associated with inductance?
Why is inductance so effective in opposing the flow of alternating
  current?

Capacitance.
Condensers are usually associated with two conducting plates
  separated by an insulator or dielectric, but actually all con-
  ductors have capacitance.
How does a condenser "store" energy?
  Energy is associated with changes in the charged particles
    which compose the atoms of the dielectric, when the plates
    are electrically charged.
  Energy stored $= (1/2)CV^2$ C in <u>farads</u>.
  Capacitance vs. elasticity in mechanical systems; C analogous
    to $1/k$.
Condenser "size."
  Capacitance $\sim \dfrac{\text{Area of plates}}{\text{Dist. betw. plates}} \times$ Dielectric const.

TENTH WEEK

B. CONTINUED
- Units - expressed in <u>farads</u>.
  - The micro-farad (mfd.), 10^{-6} farad is a more practical unit, or even micro-micro-farad.
- Dielectric constants.
  - 1 approx. for air; 5 to 10 for glass; around 2.2 for paraffin.
- Current flow through a condenser.
  - DC: A momentary charging current flows, but no further flow.
  - AC: The continual charging and discharging current, as the voltage reverses, gives effect of a continual AC current flow.
  $$I = V(2\pi f C)$$

Flow of current in AC circuits with L, C, and R (inductance, capacitance, and resistance).
- Analogous to DC circuits.
  $$I = V/Z \quad (Z = \text{"impedance"})$$
  $$= V/\sqrt{R^2 + (2\pi f L - 1/2\pi f C)^2}$$
- The quantity Z is expressed in ohms, and is analogous to resistance in DC circuits.

Resonance in circuits with L, C and R
- For the particular frequency when $2\pi f L = 1/2\pi f C$ i.e., when $f = 1/2\pi\sqrt{LC}$, the impedance Z reduces to the DC resistance, R.
- The current I becomes very large in general.
- This frequency is the natural vibration frequency of the circuit containing L, analogous to mass, and C, analogous to 1/elastic constant.
- The basis of <u>electrical oscillations</u> in radio, television, and much of our communications discussed later.

Why are these factors of far greater importance in AC circuits than in DC circuits?
- In wire communication lines, transoceanic cables, power transmission lines, radio, etc.?

Which of the experiments mentioned here would you call purely statistical, and which individual?

IN THE LABORATORY

<u>The Individual Particle.</u>
- Study the distribution in number of the cosmic rays which are received per minute in your laboratory, by means of a cosmic ray counter.
- The origin and nature of the cosmic rays is controversial at present, but they appear to be very high energy electrons, or positively charged particles, or possibly photons, coming to the earth from interstellar space. Some have energies as high as that which an electron would have if accelerated by a potential of 10 billion volts.
- The entry of a single particle into the so-called Geiger-Mueller tube counter is detected because the potential applied between the wire and cylinder of the counter is so near the sparking potential (1200-

TENTH WEEK

1600 volts) at 7 cm Hg pressure, that the few ions produced by the particle entering the chamber cause ionization by impact, and build up a tiny discharge which may be amplified and made to operate a counter. The resistance in series with the counter is about 10^9 or 10^{10} ohms, so high that the discharge cannot be maintained.

Analyze the data statistically, recording the number of particles which arrive in some small time interval, and see how nearly the results approach the expected types of distribution curve.

   Can we expect a "perfect" distribution curve with the small numbers of individual particles studied?

Is this a micro- or macro-scopic study?

What other types of phenomena would you expect to follow the same type of behavior?

Discharge in Gases.

Study the character of the discharge occurring in a discharge tube as the pressure is lowered by means of a vacuum pump. Use a high resistance in series with the high potential DC applied to the electrodes, to limit the current flow.

At what pressure does the discharge begin?

How does the discharge change as the pressure is lowered?

   What are the positions of the cathode glow, Faraday dark space, the striations, and the positive column?

If a high speed molecular diffusion pump is available, follow the character of the discharge until the pressure becomes so low it ceases. Why does it cease?

Modern Cathode Ray Tubes.

Study the operation of a modern cathode ray tube.

Note the hot oxide coated cathode source of electrons; the first grid to control intensity, the first anode whose voltage and aperture size control the focusing of the electron beam to a narrow pencil, similar to a lens in optical systems; and the accelerating anode, whose high + voltage accelerates the electrons to a high velocity, so that they have sufficient energy to cause fluorescence of the screen material.

Note the arrangement of the two pairs of electrostatic deflection plates, one pair of plates to deflect horizontally, and one pair vertically. The "linear sweep" system applies an AC voltage of "saw-tooth" wave form to the horizontal plates, so that with the proper sweep frequency, recurrent phenomena may be studied.

Study its use to observe the wave forms of alternating current, and its use with microphone and amplifier for sound wave analysis, and for other transient phenomena.

FOR STUDY AND READING

ESSENTIAL
   Lemon, pp. 268, 293-307, also 254-262.
   Or Loeb and Adams, pp. 451-455, 463-479; review pp. 334-340, 373-379, 386, 463.

TENTH WEEK

SUGGESTED
- Buckley, "History of Physics," p. 187, Historical Development of Atomic Theory of Electricity.
- Millikan, "The Electron" or "Electrons, Protons and Cosmic Rays."
- Schonland, "Atmospheric Electricity."
- Webster, Farwell, and Drew, pp. 517-534, Ions and Electrons; pp. 491-494, 497-502, Inductance and Capacity.

Eleventh Week

SECTION IV - RADIATION (Continued)

B. **THE ELECTRON** (Continued)

The Photoelectric Effect.
  What experiments indicate a connection between light and electrical phenomena?
    Were any of these known before the discovery of the electron?
  Hallwachs in 1888 discovered that a zinc plate, when illuminated by ultra-violet light, would lose its charge if charged negatively, but not if charged positively.
    How did this explain the earlier difficulties experienced by Hertz, when ultra-violet light on spark gaps caused them to break down at subnormal voltages?
    How do we know that these particles ejected from materials are really electrons?
  Effect of intensity of light.
    The number of photoelectrons ejected is directly proportional to the intensity of the light.
  Effect of frequency (or wave length) of light.
    All substances emit photoelectrons if the light has sufficiently high frequency (short wave length).
    Threshold: the minimum frequency of light which will remove photoelectrons from a given material.
  The puzzle of the connection between the energy of individual photoelectrons and the frequency.
    The energy of the individual photoelectrons ejected, above the threshold, increases with the frequency.
    But, the intensity of the light has nothing to do with the energy of the individual photoelectrons, as though each electron removal were a separate process.
    Why is this impossible to explain on "classical" grounds?
The quantum theory.
  The particle theory seems necessary to explain the photoelectric effect.
  Originally proposed by Max Planck to explain radiation from hot bodies, as we shall see.
  Light energy is composed of individual units, whose energy is proportional to the frequency.

$$E = h\nu$$

  E = energy (ergs)
  ν = light frequency
  h = Planck's constant
  6.55×10^{-27} erg.sec

Einstein photoelectric law.
  The energy of an ejected photoelectron then will be the energy of the light quantum, $h\nu$, minus the energy required to remove the electron from the atom, w_0.
    Energy photoelectron = $h\nu - w_0$
  For the alkali metals, such as sodium, potassium, etc. the energy required to remove an electron is so small that a light quantum with low energy can remove the electron, and these substances respond well to visible or even infra-red light.

ELEVENTH WEEK

B. CONTINUED

Detection of Single Photoelectrons.
  Ion counters of the type used to detect cosmic rays also detect single photoelectrons ejected from their walls or gas.
Photo-cells.
  Vacuum and inert gas-filled types.
    Emitting surfaces usually of the alkali metals prepared with great care.
    Anode with + voltage applied to attract electrons.
    Gas-filled types more sensitive since each photoelectron produces ionization by impact, increasing the total current.
    Respond almost instantaneously to light.
  Non-vacuum types.
    The Weston cell types.
      Highly sensitive, require no external battery.
      Probably based on electron interchange at surfaces of thin films under light action.
  Application of Photo-cells.
    Wide uses for "electric eye" in modern life.
    Television, picture transmission, talking movies, sorting, color grading, various industrial control devices, door openers, automatic light switches, light intensity meters, stellar photometry, etc.
    New uses are constantly being found.

Thermionic Emission.

Evaporation of electrons from hot metals - roughly analogous to vaporization of hot liquids.
Early observations on conductivity of flames, and regions near hot bodies.
Edison Effect.
  Thomas Edison in 1885 observed that current would flow from the hot carbon filament of an incandescent lamp to a wire sealed in the side of the bulb if the wire were charged positive.
The work of Richardson, Langmuir and others.
  Saturation current.
    Occurs when the positive plate voltage attracting the negative electrons from the hot cathode is large enough to attract all the emitted electrons to the plate.
  Why has the development of good vacuum pumps, "out-gassing," use of liquid air traps, and other high vacuum techniques played such an important role in this field?
  Why are such elements as tungsten or platinum used for filaments?
  How does the number of electrons emitted depend on temperature?
    Richardson's law.
$$I = A T^{\frac{1}{2}} \epsilon^{-b/T}$$
    Exponential increase with absolute temperature.
    Thermionic work functions - the energy necessary to remove an electron from the surface of a material, analogous to the same phenomenon in the photoelectric effect.

ELEVENTH WEEK

B. CONTINUED

Space charge.
> The mutual repulsion of negative electrons in the dense cloud around the filament reduces the electron flow to the plate.

Special emitting surfaces.
> Thoriated tungsten filaments. Tungsten containing thorium.
> Barium-strontium oxide-coated filaments.
>> These show greatly increased emission at lower temperatures.
>
> Indirectly heated cathodes.
>> Oxide-coated cylinders heated indirectly by hot filament inside.
>> Widely used in radio tubes for AC operation.

Two Element Tube - Plate and filament (cathode).
> Rectifier circuits.
>> Conversion of AC to pulsating DC, since current flows through tube <u>only</u> when plate is positive.
>> Half-wave and full-wave rectification.
>> High voltage rectifiers - step-up transformers.
>>> X-rays, radio transmission, nuclear disintegration.
>
> Filter circuits.
>> Use of energy stored in inductance and condensers to smooth out pulsating DC to steady DC.
>> Radio set power units.
>
> Mercury-vapor filled rectifier tubes - high efficiency.

Three Element Tubes - triodes.
> DeForest: Use of an interposed grid electrode whose - or + potential serves to control the flow of electrons from filament to plate.
>
> <u>Amplification</u>.
>> Large controlling effect on electron flow of the grid compared to the plate.
>> Characteristic curves - change of plate current with grid voltage.
>>> Amplification constant - μ
>>> = Change in plate voltage / change in grid voltage required to produce same effect on plate current.
>>> μ may be used as high as 40 or 50, 3-15 more common.
>>> Actual amplification in circuit always less.

Multi-element tubes - tetrodes, pentodes, etc.
> Extra electrodes inserted to give desired change in tube characteristics.

All-metal tubes - note the metal-glass seals.

Gas-filled tubes - filled with mercury-vapor or inert gas.
> Thyratrons - three-element tubes, with high efficiency to control large amounts of power.

ELEVENTH WEEK

B. CONTINUED

Water-cooled tubes - for high power.
Power = VI (plate voltage x electron current).
Plates must be water cooled to dissipate heat produced in stopping high energy electrons.

The electron tube as a tool.
Probably one of the most valuable and sensitive tools for scientific research; it is indispensable in the field of communications - radio, television, telephone, etc., and is becoming still more important perhaps in industrial operation and control.
Often used in conjunction with photo-cells, it makes possible almost "thinking" machines.
The most sensitive detector of small voltages and currents, may also become the basis for transmission of electrical energy at high voltage DC.

Electromagnetic Radiation.
What is the historical background for James Clerk Maxwell's prediction of electromagnetic waves, and the development of an electromagnetic theory of light and radiation?
The prediction of an electromagnetic pulse propagated throughout space when a charged particle is accelerated.
Velocity of propagation - about 186,000 miles/sec. About 2.998×10^{10} cm/sec from Michelson's measurements.
Wave length and frequency relations.
Velocity of propagation = frequency x wave length.
Hertz - 1887.
Experimental verification of electromagnetic waves predicted by Maxwell.

Electrical Oscillations (See last week).
Electrical circuits consisting of inductance (coil) and capacitance (condenser), have a natural vibration period similar to mechanical systems, since L is analogous to mass W/g, and C is analogous to 1/elastic constant, $1/k$.
Natural Frequency.
$f = 1/2\pi\sqrt{LC}$ analogous to $f = 1/2\pi\sqrt{W/gk}$
Almost any desired frequency from nearly zero to nearly infinity may be produced by changing L and C.
Resonance.
Resonance occurs between coupled electrical circuits if their natural frequencies are the same.
Damped Oscillations.
Resistance in electrical circuit causes oscillations to die out, similar to friction in mechanical circuits.
Continuous Oscillations.
If enough energy is supplied to make up for the losses due to resistance, the amplitude of the oscillations stays constant.
Vacuum tubes, properly coupled to L and C circuits do this very effectively.

ELEVENTH WEEK

B. CONTINUED
Radio.
When suitably connected to radiating systems or antennas, oscillating electrical systems radiate electromagnetic waves.
- Marconi - extension from Hertz's early work to longer wave lengths, higher power, and greater distances.
- Early Forms; Spark Transmitters.
    - Simple resonant circuit receivers, coherers, and crystal detectors.
- Later Developments.
    - DeForest - Three-element tubes.
    - Armstrong - regeneration - super-heterodyne.
- Modern Radio Transmission.
    - Wave lengths or frequencies used.
        - 600 to 10,000 meters - ships and formerly trans-atlantic. 500 to 30 KC.
        - 550-200 meters, 550-1500 KC, broadcast band.
        - 200-5 meters, 1500 KC-60 MegaC. Amateur, trans-atlantic, police, etc..
        - Experimentally down to few cm wave lengths.
- Transmitters.
    - Vacuum tube oscillators - water cooled.
    - Appropriate L and C for frequency desired.
    - Antenna length appropriate to frequency used.
    - Direction or reflecting antennas for high frequency.
- Modulation.
    - Varying amplitude of wave or "carrier" to follow waveform of voice, music, etc.
- Detection of Radio Waves.
    - Antenna - resonant tuning by coil L and condenser C.
    - Radio frequency amplification.
    - Super-heterodyne.
        - "Beating" of incoming signal with local oscillator to convert signal to frequency more easily amplified.
    - Detection
        - Rectification of modulated radio frequency voltage to pulsating DC.
    - Audio amplification.
        - Amplification of pulsating DC audio frequency voltage (40 to 7500 cycles), to operate loud-speakers, etc.

The role of the electron in the production of the complete electromagnetic spectrum.
   The range of frequencies beyond the usual radio limits.
      As we shall see, all kinds of radiation, from the longest radio waves, through the infra-red, visible, light, ultra-violet, x-ray - to the shortest gamma ray, and possibly cosmic rays, all are the same type, differing only in frequency.

IN THE LABORATORY

Photoelectric Cells.
Investigate the operation of a photoelectric cell of either the high vacuum or gas-filled types, using a sensitive suspension type galvanometer to measure the photoelectron current.

ELEVENTH WEEK

How does the photoelectron current flowing to the + charged anode depend on the applied voltage?
Can you reach "saturation current," when all the electrons are collected?

After darkening the room, use the cell to measure the intensity of illumination received at various distances from a light source.
Compare the results with an illuminometer of the photometric screen type.
Does the number of photoelectrons ejected from the cathode vary directly with the intensity of illumination in accord with the inverse square law?

If a spectrometer, monochromator, or light filters are available, investigate the response of the cell to various frequencies of light.

Two-Element Vacuum Tubes.
Investigate the characteristics of a two-element vacuum tube, containing a tungsten filament and plate.
Study the dependence of the electron emission from the filament on the filament current (temperature) and plot I_p versus I_f curves.
Study the dependence of the electron current to the plate on the plate voltage, and plot I_p versus V_p curves.
What do we mean by saturation?

Three Element Tubes.
Investigate the characteristics of three element tubes, and the way in which they amplify.
Note the internal construction in the various torn down tubes.
Study the way the electron current to the plate depends on the voltage applied to the interposed grid.
Plot an I_p versus V_g curve.
Amplification constant.
How much change in plate voltage is required to produce the same change in plate current as a one volt change in grid voltage?
This factor is the amplification constant.

Electron Emitters.
If time is available, investigate the ratio of electron emission to filament power input, for pure tungsten filaments, thoriated tungsten filaments, and barium-strontium oxide-coated filaments.
Compare the operating temperatures also.
Why are oxide coated filaments so widely used?

FOR STUDY AND READING

ESSENTIAL
Lemon, pp. 283-292, 301-310, 385-397.
Loeb and Adams, pp. 399-403, 470, 530-531.
Jauncey, pp. 182-195, 197-210.

SUGGESTED
Felix, "Television."
Henney, "Principles of Radio."
Knowlton, "Transportation of Information," pp. 396-414.
Koller, "Physics of Electron Tubes."
Lenard, "Great Men of Science," on Maxwell and Hertz.
Morecroft, "Principles of Radio Communication."

Twelfth Week

SECTION IV - RADIATION (Continued)

C. TEMPERATURE RADIATION

Radiation from Hot Bodies.
From kinetic theory, the average kinetic energy of the molecules in a material is proportional to absolute temperature.
Since molecules consist of electrical charges, electromagnetic radiation constantly occurs from all bodies, the energy radiated approaching zero at absolute zero.

Stefan-Boltzman Law.
The amount of energy radiated is proportional to the fourth power of the absolute temperature.
$$E = \sigma T^4 \qquad \sigma = \text{a constant}$$
Does it hold experimentally?

What is Meant by "Black Body Radiation"?
An ideally black body absorbs all the radiation incident on it.
It is likewise a perfect radiator.
Conversely a highly polished body reflects the incident radiation, and itself is a poor radiator.
Why are the steam radiators in this building painted black or brown? What about using aluminum paint?

Light.
If the temperature of a body is sufficiently high, some of the radiation emitted will have frequencies which are high enough to act on the photo-chemical substances in the eye and therefore produce visible light.
500°C produces a barely visible red.
What do we mean by "color"? by "white light"?
Should we limit the term "light" to apply only to visible radiation, on a purely physiological basis, i.e., the region between about 7000 A.U. (red) and 4000 A.U. (violet) wave length?
1 A.U. (angstrom unit) = 10^{-8} cm.
What about ultra-violet or infra-red light?
Do all eyes respond the same to light?
   Are the eyes good instruments for measuring intensity? Frequency?

Spectral Distribution of Energy.
Hot bodies emit continuous radiation of all frequencies or colors, tapering off toward the violet.
What happens to the color of a hot body as the temperature is raised?
Studies of the amount of energy radiated at the various frequencies (or wave lengths) show that the maximum shifts toward higher frequencies, i.e., toward the violet, as temperature is raised.
How do such light sources as the sun, about 6000°C, the tungsten electric lamps, 2000-2500°C, or carbon arcs, about 4000°C,

TWELFTH WEEK

C. CONTINUED
differ in their emission of ultra-violet, visible, and infra-red radiation?

Stellar Temperatures.
"Blue-white" stars - high temperature.
Spectral energy distribution curves indicate temperature.

Quantum Theory.
Great difficulties were experienced by Wien, Jeans, Rayleigh, and others in explaining energy distribution curves on classical grounds.
Max Planck, by introducing the concept that light energy is emitted or absorbed only in elementary units of light energy, called <u>quanta</u>, whose energy is expressed by

$$E = h\nu \qquad \begin{array}{l} E = \text{energy} \\ \nu = \text{frequency} \\ h = \text{Planck's constant} \\ 6.55 \times 10^{-27} \text{ erg.sec} \end{array}$$

explained temperature radiation satisfactorily.
We have already seen its application to the photoelectric effect.

D. RADIATION FROM EXCITED ATOMS

The radiations emitted from excited (or ionized) atoms in a discharge tube are in general sharp <u>discrete lines</u>, <u>not continuous</u>.
Why should these characteristic frequencies emitted by the various atoms be so important in leading us to understand atomic structure?

Spectroscopy. Spectral Lines. Spectral Series.
How have the prism spectrometers, the ruled gratings, the interferometer, and other tools of the spectroscopist been used over a long period to accumulate a large body of information about atomic spectra, wave lengths, etc.

Spectral Series.
How does the grouping of many lines into a series suggest a definite structure to the atom?

Rutherford-Bohr Atom Model.
Review of our early ideas of atomic structure.
How did Nils Bohr make use of Planck's quantum hypothesis to explain characteristic frequencies emitted by excited atoms?
The simple hydrogen atom - Bohr picture.
1 nuclear proton and 1 external electron.
Radius of normal electron orbit (unexcited) determined by balance between electrical attraction $F = Zee'/x^2$, and centrifugal force $F = W/g(v^2/x)$. $x =$ about $.53 \times 10^{-8}$ cm.
Quantized Orbits.
Only certain additional orbits or energy levels are possible for electron, if electron absorbs energy through excitation, <u>not just any orbit</u>.
Emission Radiation.
Electron returns to normal level in one or more transitions, not continuously. It radiates light of frequency determined

TWELFTH WEEK

D. CONTINUED

by energy given up in a transition.
$$E_1 - E_0 = h\nu$$
This picture accounts very well for the several series of lines emitted by $_1H^1$ atom. The Balmer, Lyman, Paschen, Brackett series, etc.

The New Quantum Theory.
  Further changes have been made in the original Bohr theory to account for the spectrum of helium and the more complex atoms.
  Wave Mechanics.
    The work of Schroedinger, Heisenberg, and others has shifted the basis to a more mathematical model.
    Based on a wave theory of matter, and consideration of probability and uncertainty principle.
    Energy states of atom, rather than electron itself.

Absorption and Emission of Radiation.
  The Fraunhofer Lines.
    The interior of the sun is a hot body radiating a continuous spectrum.
    Black lines in spectrum due to absorption of their characteristic frequencies by colder gases in the outer atmosphere of the sun, as the light of continuous spectrum passes through these colder gases on its way to us: i.e., they are absorption lines.
  "Resonance" in Atomic Systems.
    If the quantum falling on an atom has a frequency equal to a natural frequency of the atom, there is a high probability of the quantum being absorbed. The same frequency or a lower frequency is usually emitted in all directions.

Biological and Chemical Effects of Radiation.
  Are chemical reactions influenced by radiation?
  How much are living organisms dependent on radiation?
  Is the sun indispensable to life?
  Science A2 and B2 will consider these problems more fully.

IN THE LABORATORY

As a background for the study of the spectrometer, investigate the behavior of lenses, lens systems and prisms, by means of a "smoke box" to show the path of the light rays.
  Determine the principal focus of several converging lenses, i.e., the point at which parallel rays are brought to a focus.
  Note the production of parallel light if a point source of light is placed at focus of converging lens.
  Study in the smoke box, the functioning of a two-lens system similar to an astronomical telescope.
  Observe the deviation of light rays by a prism in the smoke box.

Prism Spectrometer.
  Investigate the construction and use of a prism spectrometer.

TWELFTH WEEK

Note the <u>collimator</u> with its slit placed at the focus of a converging lens, to produce parallel light; the prism to separate the various light frequencies (since the index of refraction depends on frequency); and the telescope with its converging lens, cross hairs, and eyepiece. Note the finely divided circles with verniers used to measure accurately the angular deviation, and hence the frequency of the light components.

Continuous Spectrum of Temperature Radiation.

Examine the continuous spectrum from a tungsten filament lamp as the filament temperature is raised from the lowest visible point, about 500°C, to the maximum.

Why does the radiation appear largely "red" at low temperatures, and show a greater content of high frequency blue and violet light at high temperatures?

Why do the older carbon filament type lamps give off a "red" light while the new tungsten lamps give off a more nearly "white" light?

How much of the total energy radiated from the filament lies below the visible, in the infra-red or heat region?

Examine the characteristic spectral lines emitted by a number of types of atoms when they are excited:
Hydrogen (note the Balmer lines)
Helium, lithium, neon, argon, cadmium, etc.
Mercury arc; sodium, the D lines, a doublet very close together.
Why does each atom have a characteristic spectrum?

If sufficient time is available, study the black Fraunhofer lines in the sun's radiation, due to absorption of their characteristic frequencies by atoms in the outer and colder part of the sun.
Prove that the black absorption line in the yellow region has the same frequency as the bright line from a sodium flame here on earth.

Visit the Ernest Kempton Adams Precision Laboratory.

Study the types of precision instruments used in the various fields of physics.

Be sure to see the Hilger quartz spectrograph, and the comparators used to measure spectral lines on plates; the infra-red spectrometer, the Millikan oil drop apparatus, etc.

FOR STUDY AND READING

ESSENTIAL
Bazzoni, "Energy and Matter," pp. 49-72.
And Lemon, pp. 347-354, 398-404, 415-428, 432-433.
Or Loeb and Adams, pp. 418-424, 432-434, 437-440, 444, 526.

SUGGESTED
Bragg, "The Universe of Light."
Dampier Whetham, "A History of Science," ch. IX, A New Era in Physics.
Knowlton, "The Electro-magnetic Spectrum," Radiation, pp. 586-593;
   Atomic Structure, pp. 599-605.
Luckiesh, "Color and Its Applications."

Thirteenth Week

SECTION IV - RADIATION (Continued)

E. X-RAYS

What is the story back of Roentgen's discovery of x-rays in 1895 through their effects on fluorescent materials?
  Would you call this discovery entirely accidental?
  What part did this discovery play in ushering in the era of modern physics?
  Why was their high penetrating power, ionizing power, later associated with waves?

Production of X-Rays.
  Bombardment of various targets (anodes) with high speed electrons.
  Why should we expect an electromagnetic wave radiation to be given off?
  Early gas-type x-ray tubes.
    Relied on ionized residual gas to produce electrons.
  Coolidge x-ray tubes.
    Hot filament sources of electrons.
    High vacuum, .0000001 mm Hg or better.
    Water cooled targets.

Nature of X-Rays.
  Are they **waves** or **particles**?
    They cause interference, characteristic of waves.
  Penetration, or conversely, absorption.
    The absorption is due largely to their giving up energy in the ejection of photoelectrons from materials.
      A quantum phenomena.
        Why should the absorption increase with density of substance? Is w/v related to number of electrons present to absorb x-rays?

Ionization by X-Rays.
  Due largely to ejection of photoelectrons from materials with considerable energy.

Wave Length or Frequency of X-Rays.
  Is it fortuitous that their maximum frequency is determined by the energy of the bombarding electrons, according to quantum law: $E = h\nu$?

The Continuous X-Ray Spectrum.
  Note how similar their energy distribution curves are to blackbody temperature radiation distribution curves.
  X-ray regions extend from about 10^{-9} cm to perhaps 5×10^{-5} cm in the ultra-violet regions, but there is no definite limit.

X-Rays and Structure of Matter.
  How did von Laue's suggestion, the use of crystals as interference gratings for x-rays, become a powerful tool to investigate crystal structure?

THIRTEENTH WEEK

E. CONTINUED

How does the chemist, the metallurgist, the crystallographer determine crystal structure from "Laue Spots"?

X-Ray Spectroscopy.
  Bragg's reflection of x-rays at critical angles from the regular planes of crystals, due to interference.
  The angle of reflection serves to measure x-ray wave-lengths.
    $\lambda = 2d\sin\theta$ - where d is distance between crystal planes.
      How has this given rise to modern x-ray spectroscopy in the hands of the Braggs, Barkla, Siegbahn, Davis, Compton, and others?
  X-Ray Line Spectra.
    X-ray line spectra superimposed on the continuous spectra are very much like the discrete lines in optical spectra.
  Moseley.
      Correlation of x-ray spectral lines with atomic weight.
    Line spectra due to excitation of inner electrons, analogous to optical spectra due to outer electrons.
    X-ray studies reveal inner electron structure.
    Quantum theory and Rutherford-Bohr atom model development influenced greatly by x-ray studies.

Physical, Chemical, and Biological Effects of X-Radiation.
  Interaction of matter, living and inorganic, with radiation one of our biggest problems.
  X-Rays in Medicine.
    Diagnostic uses - for viewing bones, internal organs, etc.
    Therapeutic uses - effect on cancer, treatment, etc.
    High voltage - penetrating x-rays for deep therapy, cancer.
      The Memorial Hospital 900,000 volt x-ray installation.
  X-Rays in Industry.
    Many applications in numerous fields, from mere inspection to essential steps in a process.

Electromagnetic Radiation.
  The complete electromagnetic spectrum.
    Long radio waves, ultra short waves, infra-red or heat waves, visible light, ultra-violet light, extreme ultra-violet, or very soft x-rays, soft and hard x-rays, gamma rays from radium, possibly cosmic rays.
    How completely have we bridged the gaps which formerly existed between the various types of waves?
      Why do we feel sure that while these waves differ greatly in the feasible modes of excitation, from one end of the scale to the other, their nature is identical?
    What modes of production are practical for the various wave length regions?
  Differences in interaction with matter are due entirely to frequency differences.
  Frequencies range from nearly zero to nearly infinity.
  Wave lengths vary from nearly infinity to nearly infinitesimal, down to at least 10^{-10} cm.
  Velocity of electromagnetic waves.

THIRTEENTH WEEK

E. CONTINUED

All follow the fundamental relation:
velocity = frequency x wave length.
All have the same velocity in vacua, approximately 2.998×10^{-10} cm/sec.
   Why have such men as Michelson spent so much time measuring this value? How fundamental is it in nature?

The Ether.
Must we have a medium for the transmission of electromagnetic waves?
The Michelson-Morley experiment.
   Failure to detect an appreciable drift of earth through ether.
   D. C. Miller's experiments.
Has the relativity theory disposed of the ether?

The Particle-Wave Controversy.
The dual aspects of radiation.
   Is radiation a wave train? Is it a stream of particles or is it a combination of both?
The continuation of the Newton-Young-Huyghens controversy.
The theories: continuous versus discontinuous nature. Maxwell - Planck.
The Compton effect.
   Scattering of x-rays by matter explained by a purely mechanical collision between two particles, an x-ray quantum with energy $h\nu$ and an electron.
Which experiments are best explained by assuming that radiation is a wave motion, and on the opposite side, which experiments seem to require a particle basis?
These problems raise fundamental questions which obviously cannot be answered completely in the present stage of scientific investigation.

IN THE LABORATORY

Before coming to the laboratory, review the previous work on spectra and spectrometers.

Gratings.
Investigate the various gratings which are available, reflection gratings, ruled transmission gratings, and grating replicas.
   Interference.
      In general all interference phenomena are based on the fact that two light beams from the same source, which traverse different paths, interfere destructively if the path difference is one half wave length (or some odd multiple) and reinforce each other if the path difference is an even wave length (or some multiple).
   What is meant by grating space? By central image? First and second orders?
      Why is the order of colors reversed from that in a prism spectrum?
   Study the grating spectra of several sources, using different grating set-ups.

Use continuous spectrum from hot filament lamp, line spectra from sodium, mercury arc, neon, and argon lamps, etc.
Given the grating constant, obtain the necessary distances and calculate the wave length of sodium light.
Can you calculate the upper and lower limits of the eye's response to light?

Interference and Diffraction Phenomena.
Investigate a number of interference and diffraction experiments which are available.
These will include:
Young's interference experiment: interference between light beams from two pin holes.
Newton's rings: interference between light beams reflected from the two sides of thin films.
Lloyd's mirror: interference between light direct from a source and a portion of the same beam reflected from a plane surface.
Measurement of thickness of extremely thin films by interference methods and others.
Diffraction patterns around various objects.

Grating Spectrograph Room.
Visit the grating spectrograph room in the basement.
Why is the grating so useful in spectroscopy?
A slightly concave grating focuses spectra analogous to the action of a concave mirror.
Investigate the so-called Rowland mounting, the slit system, position of grating, and the arrangement used for mounting photographic plates at the proper point for correct focusing.
Examine the spectrum of an iron arc, and the various orders.
Why keep the room at constant temperature, and why take such elaborate precautions in mounting?

FOR STUDY AND READING

ESSENTIAL
Knowlton, pp. 523-537, 538-543, 564, 577-585.
Or Lemon, pp. 360-371, 398-414, 427-435.
Or Loeb and Adams, X-rays, pp. 451-463; Electro-magnetic Spectrum, pp. 418-424, 430, 582-590.
Bazzoni, "Energy and Matter."

SUGGESTED
Clark, "Applied X-Rays."
Dampier-Whetham, "The New Era in Physics," p. 382.
Haas, "World of Atoms," pp. 27-40, 53-59.
Planck, "The Universe in the Light of Modern Physics."

Fourteenth Week

SECTION IV - RADIATION (Continued)

F. THE NUCLEUS

The next field of exploration in physics.
Review of material on the Rutherford-Bohr atom.

Radioactivity.
Historical Development.
What part did Becquerel's discovery in 1896 of the activity of uranium play in opening up the new era in physics?
Was this discovery really "accidental"?
What is the story back of the discovery of polonium and radium by Professor and Mme. Curie?
Why do we believe that radioactivity is a <u>nuclear</u> phenomenon?
The Three Types of Radiations.
Alpha particles.
How did Rutherford's spectroscopic test prove that each was a helium atom minus two electrons?
Ionization.
Heavy ionization produced along their path.
Lose energy and eventually stop.
Range.
Characteristic, sharply defined ranges (or energies), from uranium I, 2.73 cm; polonium (RaF), 3.9 cm; RaC', 6.97; on up to ThC', 8.62 cm.
Energies range from UI, 4 MEV, to ThC', 8.76 MEV.
MEV - million electron volts. The energy which an electron would have if accelerated by 1 million volts.
Are there energy levels in the nucleus?
Some elements emit several groups of alpha particles.
Beta radiation.
High energy electrons.
Far more penetrating than alpha particles, perhaps 100 times on average.
Ionization along track much less.
Energy.
Some have nearly the velocity of light, but since mass is small, energies are usually less than 3 MEV.
Gamma radiation.
Electromagnetic waves.
Higher frequency and therefore more penetrating than x-rays.
RaC' γ radiation, 1.8 MEV, is absorbed about 50% by 1 cm of lead.
Gamma ray quanta absorbed by photoelectric ejection of electrons, Compton recoils, formation of electron and positron pairs, etc.
The Radioactive Series.
The apparent mechanical-electrical instability of a large number of the higher atomic weight atoms.

FOURTEENTH WEEK

F. CONTINUED

The three series - uranium, actinium, and thorium.
Uranium series:

$$_{92}U^{238} \rightarrow {}_{88}Ra \rightarrow {}_{86}Radon \rightarrow {}_{84}RaA \rightarrow {}_{82}RaB \rightarrow {}_{83}RaC \rightarrow$$

$$_{84}RaC' \rightarrow {}_{82}RaD \rightarrow {}_{83}RaE \rightarrow {}_{84}RaF(polonium) \rightarrow {}_{82}RaG(lead)$$

An approximately similar series for thorium and actinium.

A Complicated Series of Transformations.
The unstable parent atom emits α particles, β particles, and γ rays, involving a long series of transformations, each new atom formed by expelling a particle also being unstable, until finally the atom becomes an isotope of lead, stable at last.

The Nuclear Problem.
What do we know about the size and properties of the nucleus?
Method of Attack.
Only method of approach to this "central sun" is to shoot particles at nucleus and see what happens.
Particles Composing the Nucleus.
These particles come out of nucleus under bombardment.
Protons - $_1H^1$ nuclei, + charge, mass about 1.0078.
Neutrons - $_0n^1$, no charge, mass about 1.0089, still a matter of uncertainty. (Discovered by Chadwick, 1932.)
Protons and neutrons appear to be the basic particles of which the nucleus is constructed.
Perhaps these may also be complex.
Other Nuclear Units.
Alpha particles - $_2He^4$ nuclei, + + charge, mass 4.0022.
Probably composed of 2 protons and 2 neutrons.
Deuterons - $_1D^2$, hydrogen isotopes.
Probably a proton + neutron.
Likewise $_1H^3$ hydrogen isotope and $_2He^3$ helium isotope.
Electrons and Positrons.
Probably do not exist in nucleus as such.
Electrons, neutron conversion to proton?
Positrons (positive electrons).
Discovered by Anderson in cosmic ray cloud chamber studies.
Annihilated quickly by combining with electron, producing γ radiation, i.e., conversion of matter into radiation.
Conversely produced by γ radiation conversion to positron and electron pair, i.e., conversion of radiation into matter.
Conversion of proton to neutron in nucleus may also produce a positron?

Equivalence of Matter and Energy.
Apparently radiation may be converted into matter and matter into radiation.
Relativity relation.
Energy associated with matter
$E = mc^2$ where c is velocity of light
$m = W/g$

FOURTEENTH WEEK

F. CONTINUED

Energy of 1 gm matter ≃ about 10^{18} gm.cm.
The energy of the universe is in the nucleus, where the mass is.

The Nucleus as a Unit.
Why doesn't the nucleus fly apart if inverse square law of repulsion exists between + charged particles in nucleus?
At some point, close in, the repulsive forces must change to attraction.
Neutrons as the cement which holds together?
Nucleus as a potential hill.
> A charged particle must work against an increasing repulsive force as it approaches nucleus, but if it has sufficient energy may climb over hill and get in.

Producing Nuclear Projectiles.
Natural Radioactive Materials.
> The first bombarding particles - Rutherford.
> Alpha particles from various radioactive elements, polonium, (RaF), radon.

High Voltage Methods.
> Acceleration of protons, deuterons to high energies in vacuum tubes by high potentials.
>> Cockcroft and Walton.
>> The van de Graaf generator, 1,000,000 volts and more.
> Acceleration in small steps.
>> The "cyclotron" - Lawrence.
> Acceleration to high energy by many stages.

Artificial Nuclear Transformations.
How does the modern alchemist produce his results?
The work of Rutherford, Chadwick, Curie-Joliot, and others.
Three General Transmutation Processes.
> When a particle enters nucleus, three general processes can occur.
>> A new stable nucleus may be formed.
>> Other heavy particles, neutrons, protons, deuterons, alpha particles, may be immediately expelled from the nucleus to make a new stable nucleus.
>> An unstable, <u>artificially radioactive</u>, nucleus may be formed, which eventually becomes stable by expelling an electron or positron at some later time.
>> Gamma radiation often accompanies these nuclear reorganizations.

Two Fundamental Problems.
> + charged particles - protons, deuterons, alpha particles, etc. experience enormous repulsive forces as they approach + charged nucleus.
>> Energies of order of 3 MEV or more are required to allow particle to climb over potential hill into nucleus.
> The effective nuclear "size" is so small (10^{-13} to 10^{-12} in radius, compared to about 10^{-8} cm or more for atom as a whole) that only a very small proportion of projectiles would strike nucleus squarely, even if they had enough energy.

FOURTEENTH WEEK

F. CONTINUED

Particle-wave Theory, Resonance.
Since particles behave like waves, there is always some amplitude inside nucleus, even though extremely small. Hence, even at voltages as low as 50,000, particles have a finite though extremely small probability of entering nucleus. Resonance also occurs such that at certain special energies, a particle penetrates into nucleus very easily.

Typical Nuclear Reactions.
Alpha particles from radioactive materials

$$_2He^4 + Energy + {_4Be^9} \rightarrow {_6C^{12}} + {_0n^1} + Energy$$

Mass converted into energy.

Protons accelerated by high voltages

$$_1H^1 + Energy + {_3Li^6} \rightarrow {_2He^3} + {_2He^4} + Energy$$

Deuterons accelerated by high voltages

$$_1D^2 + Energy + {_3Li^6} \rightarrow {_3Li^7} + {_1H^1} + Energy$$

Careful measurements of atomic masses by mass spectrographs by Aston, Bainbridge, and others appears to check very closely the mass-energy balance in nuclear transformations.

Artificially Radioactive Nuclei.
In many cases the atom produced is not entirely stable, and in turn disintegrates giving off an electron or positron and often γ radiation, becoming stable at last.

$$_1D^2 + {_6C^{12}} + Energy \rightarrow {_7N^{13}} + {_0n^1} + Energy$$

$$_7N^{13} \rightarrow {_6C^{13}} + {_1e^+}$$

γ radiation also accompanies many reactions and must be added to make equations balance.

Interaction of Neutrons with Matter.
Uncharged - its properties are quite different from other particles.
Our best nuclear projectile.
Fast Neutrons.
  Interact only with nuclei.
  No repulsive force since uncharged.
  Largely elastic collisions with nuclei, like billard ball collisions.
  An ideal probe to measure nuclear "sizes."
Slow Neutrons.
  When slowed down by impacts with H nuclei in paraffin from 1 - 14 MEV down to ordinary thermal energies, 1/40 EV, properties change enormously.

FOURTEENTH WEEK

F. CONTINUED

Enormously increased nuclear transformations.
    Being uncharged, enter practically all nuclei easily, and produce transformations.

Large nuclear interaction.
    Some nuclei, such as cadmium, gadolinium, interact many thousands of times greater with slow neutrons.

Reactions analogous to combustion in neutron gas

$$_0n^1 + {}_3Li^6 \rightarrow {}_2He^4 + {}_1H^3 + 2 \text{ or } 3 \text{ MEV}$$

Liberation of atomic energy.
    Will such reactions become practical sources of energy? Can we drive ships across the ocean with 1 gm of Li?

Nuclear Structure.
Out of this mass of rapidly accumulating experimental data, in time a theory of nuclear structure can be built.

Matter, Energy, and Radiation.
Does then the fact that matter may be transformed into energy, that pure radiation may be changed into matter, etc., mean that these are so closely related that they are equivalent?

The physicist with the help of the scientists in other fields is beginning to attack these fundamental questions. A great amount of knowledge has already been accumulated, but the surface has only been touched. The future lies ahead, to give man far greater understanding and power over the world in which he lives.

IN THE LABORATORY

Physics Research Laboratories.
In small groups under the guidance of an instructor, visit some of the physics research laboratories not already visited.

X-ray Laboratories - Professor Davis - 7th Floor.
Investigate the high voltage full wave rectifier systems, with their condenser filters, for supplying 100,000 or 300,000 volts, DC, from 500 cycle AC.

Examine the Coolidge type water-cooled, hot filament x-ray tubes used in most of the work.
    Note the high vacuum system for exhaustion of x-ray tubes before sealing off, with the high speed mercury-vapor diffusion pumps, the rotary oil fore pump, the liquid air traps, and the "baking out" oven.
    How good is an x-ray vacuum?

Study the x-ray crystal spectrometer set-ups.
    How are the crystals used to measure x-ray wave lengths?
    Note the sensitive vacuum tube electrometers used to measure the ionization currents produced by x-rays as a detector.

What sort of knowledge have such studies yielded on the nature of matter?

FOURTEENTH WEEK

The Neutron and Nuclear Disintegration.
Laboratory on 13th floor.
What protection must be used from the gamma radiation given off by the radon used to produce the neutrons from beryllium?
What types of amplifiers and counting equipment are used to detect the individual particles?
How have the processes of production of neutrons, the studies of the nature of the neutron, the studies of its interaction with atomic nuclei, and its use in producing new nuclei opened up new fields of investigation regarding the nature of matter?

Cloud Chamber and Professor Webb's Spectrographic Laboratories - 6th Floor
Investigate the mechanism of the cloud chamber for producing expansions and observe the tracks produced by particles, such as alpha particles, or some other nuclear transformation, if the apparatus is available.
How does a cloud chamber work?
Note the photographic mechanism for taking stereographic pictures.
In the spectrographic laboratories investigate the large Hilger quartz spectrographs, and the methods of measuring wave lengths of spectral lines.

The "Cyclotron" - Large Magnets - Basement.
How is the circular path of a charged particle in a magnetic field utilized in accelerating particles to a high velocity for nuclear disintegration by successive steps?
How are the accelerating voltages from the high frequency oscillator applied to the system?
Note the oscillator system, its tubes, inductance, and capacity, and the high voltage DC rectifier system.
Why are these large magnetic fields so useful in physics?
Will such methods of producing nuclear transformations supersede natural radioactive methods?

FOR STUDY AND READING

ESSENTIAL
Loeb and Adams, Radioactivity, pp. 484-509; Nuclear Transmutations, pp. 561-567, 572-582, 605-608.
Or Bazzoni, "Energy and Matter," pp. 93-115.
And Lemon, pp. 316-330.

SUGGESTED
K. K. Darrow, series of articles in "Review of Scientific Instruments," on the Nucleus and Transmutation, 1933, 1934, and 1935 volumes.
Loeb and Adams, Isotopes - Nuclear Atoms, pp. 509-527, 567-572, 601-605; Indeterminancy Principle, pp. 547-601.
Pittsburgh staff, "Atomic Physics."
Richtmeyer, "Introduction to Modern Physics."
Rutherford, Chadwick, and Ellis, "Radiation from Radioactive Substances."

The Science Reading Room

The Science Reading Room with its carefully selected collection of books has been established especially to give students an opportunity to widen their horizons and to satisfy any inclination for careful study of any phase of science which interests them. Books of all types ranging from very popular treatments to complete treatises are available. A number of scientific periodicals have been added to permit students to follow the latest developments.

In general there are no arbitrary readings in Science A1; how much the student will take advantage of the opportunity depends upon himself. The following books and periodicals in the collection are most useful in Science A1.

GENERAL
    Barnes. Scientific Theory and Religion.
    Bragg. Concerning the Nature of Things.
      Old Trades and New Knowledge.
    Bridgman. The Logic of Modern Physics.
    Buckley. History of Physics.
    Cajori. History of Physics.
    Chase. Men and Machines.
    Crew. Rise of Modern Physics.
    Dampier-Whetham. A History of Science.
    Hallock and Wade. Evolution of Weights and Measures.
    Haswell. Horology.
    Heyl. New Frontiers of Physics.
    Hodgins-Magoun. Behemoth (story of power).
    Humphreys. Physics of the Air.
    Joffe. Crucibles.
      The Physics of Crystals.
    Kaempffert. A Popular History of American Invention.
    Lenard. Great Men of Science.
    Lodge. Pioneers of Science.
    Luckiesh. Foundations of the Universe.
    Newman. Nature of World and Man.
    Planck. Universe in the Light of Modern Physics.
    Pupin. From Immigrant to Inventor.
    Soddy. Matter and Energy.
    Wyckoff. Structure of Crystals.

MODERN PHYSICS
    Andrade. Structure of the Atom.
    Aston. Mass Spectra and Isotopes.
    Berthoud. New Theories of Matter and Atoms.
    Clark. Applied X-Rays.
    Compton. X-Rays and Electrons.
    Cox. Time-Space and Atoms.
    Crowther. Ions, Electrons, and Ionizing Radiations.
    Darwin. The New Concepts of Matter.
    Eddington. Nature of the Physical World.
      Space Time and Gravitation.
      Stars and Atoms.

MODERN PHYSICS (Continued)
    Einstein. Relativity.
    Haas. The World of Atoms.
    Hughes and DuBridge. Photo-Electric Phenomena.
    Jauncey. Modern Physics.
    Jeans. Stars in Their Courses.
        The Mysterious Universe.
        The Universe Around Us.
    Kaye. X-Rays.
    Millikan. The Electron; Photons, Electrons and Cosmic Rays.
    Mills. Within the Atom.
    Pittsburgh staff. Outline of Atomic Physics.
    Richtmeyer. Introduction to Modern Physics.
    Russell. A B C of Relativity.
    Rutherford, Chadwick, and Ellis. Radiations from Radioactive Substances.

ASTRONOMY
    Fath. Elements of Astronomy.
    Hale. Signals from the Stars.
    Hutchinson. Splendour of the Heavens.
    Russell, Dugan, and Stewart. Astronomy.

ELECTRICITY AND MAGNETISM
    Felix. Television.
    Henney. Principles of Radio.
    Koller. Physics of Electron Tubes.
    Morecroft. Elements of Radio Communication.
    Pohl. Electricity and Magnetism.
    Starling. Electricity and Magnetism.

HEAT
    Edser. Heat for Advanced Students.
    Loeb. Nature of a Gas.

LIGHT
    Bragg. The Universe of Light.
    Clerc. Photography, Theory and Practice.
    Luckiesh. Color and Its Applications.
    Roebuck. The Science and Practice of Photography.

MECHANICS
    Brooks and Wilcox. Engineering Mechanics.
    Lindsay. Physical Mechanics.

SOUND
    Bragg. The World of Sound.
    Fletcher. Speech and Hearing.
    Knudsen. Architectural Acoustics.
    Miller. Science of Musical Sounds.
    Stewart. Introductory Acoustics.

TEXTBOOKS
    Kimball. College Physics.
    Knowlton. A Textbook of Physics.

THE SCIENCE READING ROOM

TEXTBOOKS (Continued)
    Lemon. From Galileo to Cosmic Rays.
    Loeb and Adams. Development of Physical Thought.
    Saunders. A Survey of Physics.
    Webster, Farwell, and Drew. General Physics.

TABLES
    Chemistry and Physics Handbook.
    Smithsonian Physical Tables.

PERIODICALS
    Nature.
    Popular Mechanics.
    Review of Scientific Instruments.
    Science Service News Letter.
    Scientific American.
    Scientific Monthly.

COLUMBIA UNIVERSITY PRESS
Columbia University
New York

FOREIGN AGENT
OXFORD UNIVERSITY PRESS
HUMPHREY MILFORD
Amen House, London, E.C. 4

Bei Fragen zur Produktsicherheit wenden Sie sich bitte an:
If you have any questions regarding product safety,
please contact:

Walter de Gruyter GmbH
Genthiner Straße 13
10785 Berlin
productsafety@degruyterbrill.com